Ben Stacy Jerrik (Ed.)

Maine State Route 35

Ben Stacy Jerrik (Ed.)

Maine State Route 35

Cumberland and Oxford Canal, U.S. Route 302, Maine State Route 5, U.S. Route 2

Part Press

Imprint

Permission is granted to copy, distribute and/or modify this document under the terms of the GNU Free Documentation License, Version 1.2 or any later version published by the Free Software Foundation; with no Invariant Sections, with the Front-Cover Texts, and with the Back- Cover Texts. A copy of the license is included in the section entitled "GNU Free Documentation License".

All parts of this book are extracted from Wikipedia, the free encyclopedia (www.wikipedia.org).

You can get detailed informations about the authors of this collection of articles at the end of this book. The editors (Ed.) of this book are no authors. They have not modified or extended the original texts.

Pictures published in this book can be under different licences than the GNU Free Documentation License. You can get detailed informations about the authors and licences of pictures at the end of this book.

The content of this book was generated collaboratively by volunteers. Please be advised that nothing found here has necessarily been reviewed by people with the expertise required to provide you with complete, accurate or reliable information. Some information in this book maybe misleading or wrong. The Publisher does not guarantee the validity of the information found here. If you need specific advice (f.e. in fields of medical, legal, financial, or risk management questions) please contact a professional who is licensed or knowledgeable in that area.

Any brand names and product names mentioned in this book are subject to trademark, brand or patent protection and are trademarks or registered trademarks of their respective holders. The use of brand names, product names, common names, trade names, product descriptions etc. even without a particular marking in this works is in no way to be construed to mean that such names may be regarded as unrestricted in respect of trademark and brand protection legislation and could thus be used by anyone.

Cover image: www.ingimage.com
Concerning the licence of the cover image please contact ingimage.

Publisher:
Part Press is a trademark of
International Book Market Service Ltd., 17 Rue Meldrum, Beau Bassin, 1713-01 Mauritius
Email: info@bookmarketservice.com
Website: www.bookmarketservice.com

Published in 2011

Printed in: U.S.A., U.K., Germany. This book was not produced in Mauritius.

ISBN: 978-613-8-94727-1

Contents

Articles

References

Maine_State_Route_35

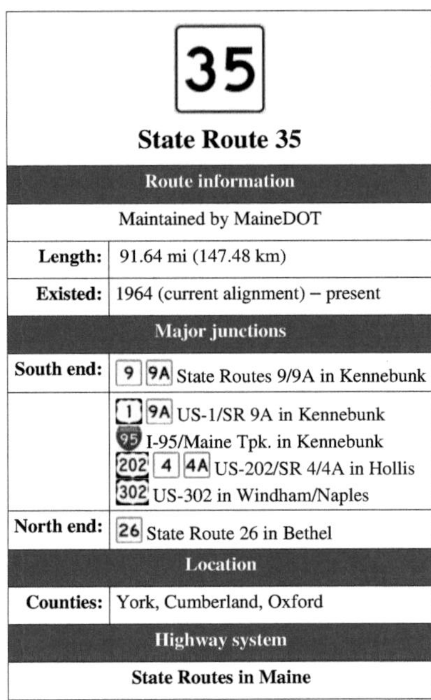

Maine State Route 35 runs the course of western Maine, from Bethel to Kennebunk. It passes through Oxford, Cumberland and York Counties. It is known in its lower sections for both its unusually windy course as well as its notoriously poor paving, as a result of winter frost heaves. Its northern section leads to the famous ski resort, Sunday River. The route crosses the Presumpscot River and a well preserved section of the Cumberland and Oxford Canal approximately one mile west of U.S. Route 302 in North Windham.

There is currently a state of confusion regarding where exactly Route 35 exists between the junction at Hunt's Corner Road and Route 5, and the town of Bethel. Historically, Route 35 leaves Route 5 at this point and takes a more easterly route toward Bethel, ending up being signed as Vernon St at the junction of Main Street in Bethel. Current (2006) maps from the American Automobile Association and Mapquest still show this route as Route 35. However, in reality, as of July 4, 2006 Route 5 is signed as both Routes 5 and 35 all the way to U.S. Route 2 in Bethel. The question of when the "old" Route 35 was switched to follow Route 5 is an open issue. Personal experience from November 1, 2004 indicated that the "old" Route 35 was barely being maintained, so one could speculate that the decision to realign this state route to double up with Route 5 happened roughly near this time.

References

Ref: Department of Transportation, Augusta, Maine. Public Records File.

Maine_Department_of_Transportation

The Maine Department of Transportation, also known as MaineDOT, is the bureaucratic office of the state government charged with the regulation and maintenance of roads and other public infrastructure in the state of Maine. MaineDOT reports on the adequacy of roads, highways, and bridges in Maine. It also monitors environmental factors that affect the motor public such as stormwater, ice/snow buildup on roads, and crashes with moose.

Organization

MaineDOT is an agency that consists of several offices:

- Bureau of Planning
- Bureau of Maintenance and Operations
- Office of Passenger Transportation
- Office of Freight Transportation
- Office of Communications
- Bureau of Project Development
- Capital Resource Management
- Transportation Service Center
- Environmental Office
- Office of Legal Services and Internal Audit
- Safety Office
- Contract Procurement Office
- Office of Engineering Quality and Oversight

External links

- MaineDOT[1]
- US Department of Transportation[2]

Sources

- Working with the MaineDOT: A Guide for Municipal Officials [3]

References

[1] http://www.maine.gov/mdot/index.php
[2] http://www.dot.gov/
[3] http://www.maine.gov/mdot/working-with-dot/pdf/2007workingwithmaineDOT.pdf

Cumberland_and_Oxford_Canal

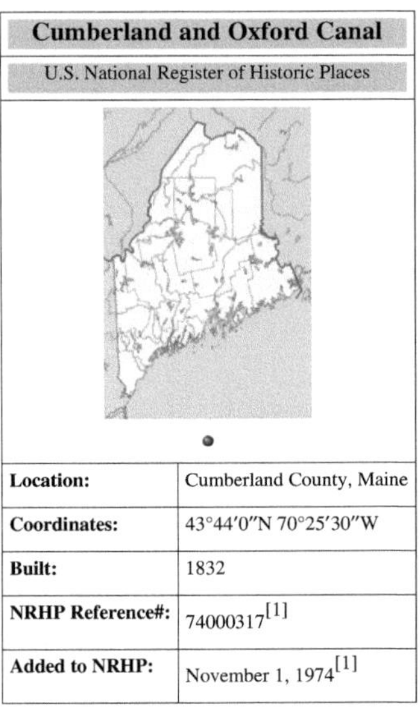

Location:	Cumberland County, Maine
Coordinates:	43°44′0″N 70°25′30″W
Built:	1832
NRHP Reference#:	74000317[1]
Added to NRHP:	November 1, 1974[1]

The **Cumberland and Oxford Canal** was opened in 1832 to connect the largest lakes of southern Maine with the seaport of Portland, Maine. The canal followed the Presumpscot River from Sebago Lake through the towns of Standish, Windham, Gorham, and Westbrook. The Canal diverged from the river at Westbrook to reach the navigable Fore River estuary and Portland Harbor. The canal required 27 locks to reach Sebago Lake at an elevation of 267 feet (81 m) above sea level. One additional lock was constructed in the Songo River to provide 5 feet (1.5 m) of additional elevation to reach Long Lake from Sebago Lake. Total navigable distance was approximately 38 miles (61 km) from Portland to Harrison at the north end of Long Lake. A proposed extension from Harrison to Bear Pond and Tom Pond in Waterford would have required three more locks on the Bear River, but they were never built.[2]

A state lottery was authorized to help raise $50,000 for the project, and the Canal Bank of Portland was chartered in 1825. The canal was completed in 1830 at a cost of $206,000.[3] The excavated portions of the canal had a surface width of 30 feet (9.1 m) with a 10 feet (3.0 m)

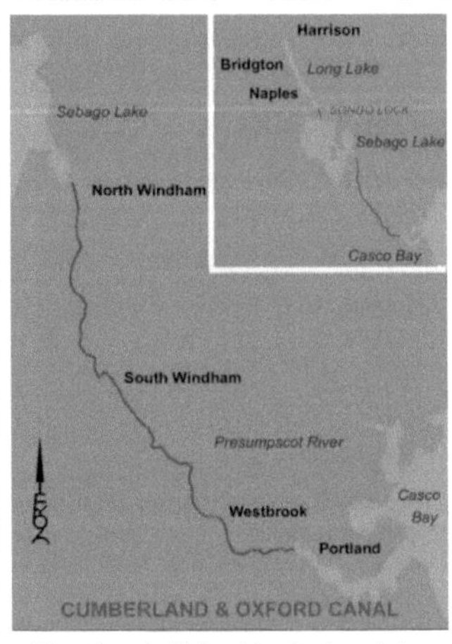

Map of canal with inset showing Long Lake, Songo Lock, and Sebago Lake

wide channel 3.5 feet (1.1 m) deep. The locks were 10 feet (3.0 m) wide and 80 feet (24 m) long. Lock walls were made of granite masonry with wooden gates at either end. A lock keeper was stationed at each lock to move the lock gates with heavy timber balance beams, manipulate iron valves to adjust water levels within the lock, and collect a 6 cent fee for use of the lock.[4]

Canal boats

The flat-bottomed canal boats had blunt bows, square sterns, and a draft of 3 feet (0.91 m). A tow path adjacent to the excavated portions of the canal enabled horses to tow the canal boats while the boatmen steered with poles. Canal boats using the lakes had a removable keel and two short hinged masts capable of supporting sails or being folded down for passage through the excavated canal. Cargos included lumber, masts, barrel hoops and staves, boxmaking shook, and firewood from the interior to Portland. Apples were an important agricultural product of the area; and Oriental Powder Company mills adjacent to the canal in Windham manufactured nearly 25% of the Union gunpowder supply for the American Civil War.[5] Canal boat passengers were charged one-half cent per mile. A wide variety of manufactured goods moved inland through the canal from Portland. The south end of Long Lake is locally known as Brandy Pond because a barrel of brandy was lost from a canal boat during passage through that part of the waterway.[6]

Railroads and Steamboats

Freight to and from Oxford County began moving over the Atlantic & St Lawrence Railroad (later Grand Trunk Railway and then Canadian National Railway Berlin Subdivision) in the 1850s and the 18-mile (29 km) long Presumpscot River portion of the canal fell into disuse when the Portland and Ogdensburg Railroad (later Maine Central Railroad Mountain Division) reached Sebago Lake Station in 1870. Some of the Presumpscot River lock facilities were converted to dams for the S. D. Warren Paper Mill. Steamboats continued to use Songo Lock to provide transportation from Sebago Lake Station to the lakeside

A steamboat of the type using the canal between Sebago Lake and Long Lake about 1910

communities of Bridgton, Harrison, Naples, Sebago, Casco, Raymond and North Windham. The Bridgton and Saco River Railroad reached Bridgton in 1883 and Harrison in 1898. The Sebago Lake, Songo River, and Bay of Naples Steamboat Company continued to offer summer passenger service to tourists until the last steamboat *Goodrich* burned at its Naples dock in 1932. Songo Lock remains in service for pleasure boats.

Relicensing of the dams was the subject of the 2006 Supreme Court case S. D. Warren Co. v. Maine Board of Environmental Protection.

See also

- National Register of Historic Places listings in Cumberland County, Maine
- Fore River Sanctuary

Notes

[1] "National Register Information System" (http://nrhp.focus.nps.gov/natreg/docs/All_Data.html). *National Register of Historic Places*. National Park Service. no date specified. .

[2] Ward, Ernest E. *My First Sixty Years in Harrison, Maine* Cardinal Printing 1967 p.7

[3] Ward, Ernest E. *My First Sixty Years in Harrison, Maine* Cardinal Printing 1967 p.8

[4] Ward, Ernest E. *My First Sixty Years in Harrison, Maine* Cardinal Printing 1967 p.9

[5] "Historic Interpretive Signs about the Oriental Powder Mills Installed" (http://www.prlt.org/PDF/2010winter_newsletter.pdf). Presumpscot Regional Land Trust. . Retrieved 2010-07-30.

[6] Ward, Ernest E. *My First Sixty Years in Harrison, Maine* Cardinal Printing 1967 p.10

References

- Johnson, Ron (undated). *Maine Central R.R. Mountain Division*. 470 Railroad Club.
- Jones, Robert C. (1993). *Two Feet to the Lakes, The Bridgton & Saco River Railroad*. Pacific Fast Mail.
- Moody, Linwood W. (1959). *The Maine Two-Footers*. Howell-North.
- Meade, Edgar T., Jr. (1968). *Busted and Still Running*. The Stephen Greene Press.
- Ward, Ernest E. (1967). *My First Sixty Years in Harrison, Maine*. Cardinal Printing.
- "National Register of Historical Places - MAINE (ME), Cumberland County" (http://www.nationalregisterofhistoricplaces.com/ME/Cumberland/vacant.html). Retrieved 2007-12-15.

U.S._Route_302

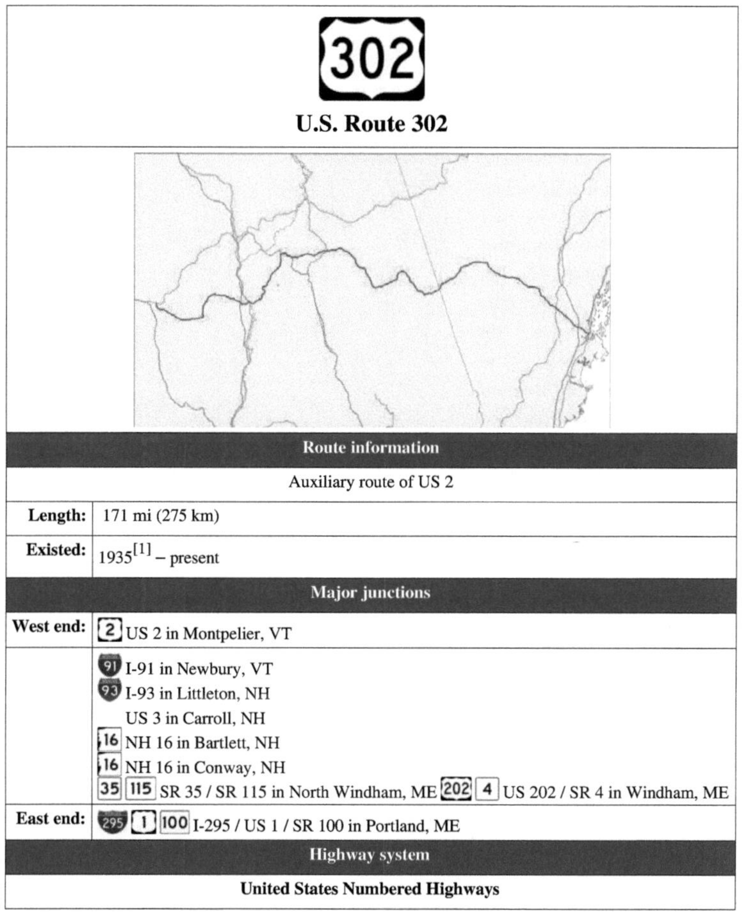

Route information		
Auxiliary route of US 2		
Length:	171 mi (275 km)	
Existed:	1935[1] – present	
Major junctions		
West end:	US 2 in Montpelier, VT	
	I-91 in Newbury, VT	
	I-93 in Littleton, NH	
	US 3 in Carroll, NH	
	NH 16 in Bartlett, NH	
	NH 16 in Conway, NH	
	SR 35 / SR 115 in North Windham, ME US 202 / SR 4 in Windham, ME	
East end:	I-295 / US 1 / SR 100 in Portland, ME	
Highway system		
United States Numbered Highways		

U.S. Route 302 (US 302) is a spur of U.S. Route 2. It currently runs 171 miles north (275 km) from Portland, Maine, at U.S. Route 1, to Montpelier, Vermont, at US 2. It passes through the states of Maine, New Hampshire and Vermont.

Route description

Maine

US 302 is known as the Roosevelt Trail through southern Maine because it was the beginning of the Theodore Roosevelt International Highway to Portland, Oregon.[2] [3] The highway leaves Portland, Maine, bridging the Presumpscot River into Westbrook at Riverton. The Roosevelt Trail then bridges the Pleasant River at milepost 13.4 in Windham, the Crooked River in Casco near a boyhood home of Nathaniel Hawthorne, Long Lake near milepost 31 in Naples, Moose Pond near milepost 46 in Bridgton, and the Saco River near milepost 56 in Fryeburg. The highway follows the Saco River from Fryeburg into the White

U.S. Route 302 in Berlin, Vermont, approximately two miles from the western terminus

Mountains and enters New Hampshire near milepost 58.[4] It is a two-lane highway for almost all of its length, but there are multi-lane sections within the Portland area, as well as short four-lane sections in and around North Windham (especially the commercial areas). Some of the hilly sections also feature a third passing lane.

New Hampshire

U.S. Route 302 enters New Hampshire following the Saco River through Crawford Notch in the White Mountains. The highway follows the Ammonoosuc River out of the mountains and bridges the Connecticut River into Wells River, Vermont.

History

The southern end of US 302 follows city streets from Longfellow Square in the 17th century colonial seaport of Portland, Maine. The highway follows a 19th century stagecoach road from Portland through Windham to Bridgton. The portion from Windham to Bridgton was built about 1785. Stagecoach service began in 1803, and the route became a post road for the United States Postal Service in 1814. Transportation over this route was augmented by the Cumberland and Oxford Canal from 1832 to 1932, and by the Bridgton and Saco River Railroad from 1883 to 1941.[5] The highway through Crawford Notch follows the Tenth New Hampshire Turnpike built in 1803 and parallels the Maine Central Railroad Mountain Division built in 1877. The highway eliminated railway passenger travel over the route from Portland by 1958 and railroad freight service through Crawford Notch was discontinued in 1983.[6]

Longfellow Square marks the southern end of U.S. Route 302.

From 1922 until 1935, much of what is now US 302 was a part of the New England Interstate road system, designated as **New England Interstate Route 18** (NE-18) from Portland, Maine, northwest to Littleton, New Hampshire (roughly 112 miles). From Littleton west to Montpelier in Vermont, US 302 and NE-18 took different paths. NE-18 took a more northerly route, along present-day New Hampshire Route 18 and Vermont Route 18 to St. Johnsbury, Vermont (closely paralleling I-93), then along present-day US 2 up to Montpelier.

Current US 302 runs along a more southerly route using other former sections of New England Interstate Routes. From Littleton, it went along former NE-10 to Woodsville, New Hampshire, then along former NE-25 to Montpelier.

See also

Bannered routes

- U.S. Route 302 Business (Bartlett, New Hampshire), a loop connecting US 302 to Lower Bartlett. It is locally known as the **Intervale Resort Loop** and is signed only as New Hampshire Route 16A.

Related state highways

- New Hampshire Route 18
- Vermont Route 18

References

[1] Droz, Robert V. U.S. Highways : From US 1 to (US 830) (http://www.us-highways.com/usbt.htm). URL accessed 27 February 2006.

[2] Tracy, A.W. *Theodore Roosevelt International Highway* (1996) p.7

[3] "U.S. 2: Houlton, Maine, to Everett, Washington" (http://www.fhwa.dot.gov/infrastructure/us2.cfm). United States Department of Transportation Federal Highway Administration. . Retrieved 2011-09-05.

[4] Greaton, Everett F. *Maine, a Guide "Down East"* (1937) pp.375-381

[5] Jones, Robert C. *Two Feet to the Lakes: The Bridgton & Saco River Railroad* (1993) ISBN 0-915713-26-8 pp.12-13

[6] Johnson, Ron *Maine Central Railroad Mountain Division* p.9

Browse numbered routes

Maine_State_Route_5

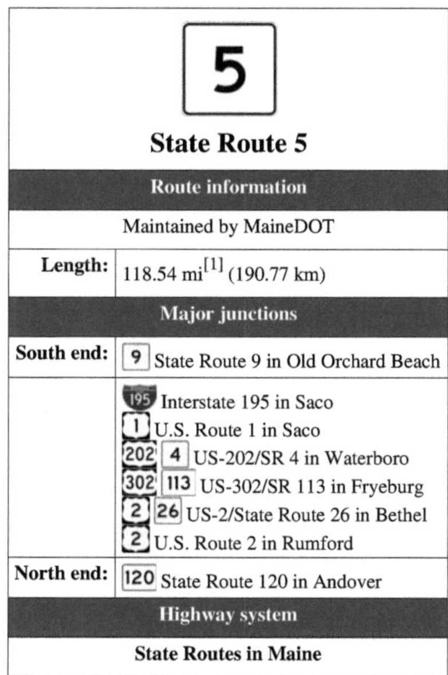

State Route 5 is part of Maine's system of numbered state highways, running from the an intersection with Route 9 in Old Orchard Beach, to an intersection with Route 120 in Andover. Route 5 is 118.1 miles (190.1 km) long.

From its southern terminus near the Pier in Old Orchard Beach, Route 5 leaves the town to the west, going towards the neighboring city of Saco. The route runs northwest from Saco, passes very briefly through a Northwest corner of Biddeford, and intersects U.S. Route 202 near the Lyman-Waterboro line. Route 5 runs concurrently with U.S. 202 for a short distance to East Waterboro.

Between Waterboro and Cornish, Route 5 is known as the **Sokokis Trail**. North of Cornish, the highway follows the Saco River, crossing it at Hiram, to the town of Fryeburg. Route 5 continues north through Lovell to Bethel, where it intersects U.S. Route 2. The two routes run together along the Androscoggin River to Rumford Point in the town of Rumford, where Route 5 leaves to the north. It follows parallel to the west bank of the Ellis River to the route's end in Andover.

Concurrent routings

- U.S. Route 202 and Route 4: 1.9 miles (3.1 km), Lyman to Waterboro
- Route 117: 6.3 miles (10.1 km), Cornish to Hiram
- Route 113: 18 miles (29 km), Baldwin to Fryeburg
- Route 35: 6.2 miles (10.0 km), Albany Township
- U.S. Route 2: 12 miles (19 km), Bethel to Rumford
- Route 26: 5.8 miles (9.3 km), Bethel to Newry

Route 5A

Route 5A is a short 3-mile (4.8 km) loop off Route 5 in Lovell.

References

[1] Maine State Route log via floodgap.com: Maine State Route 5 (http://www.floodgap.com/roadgap/me/r?5)

U.S._Route_2

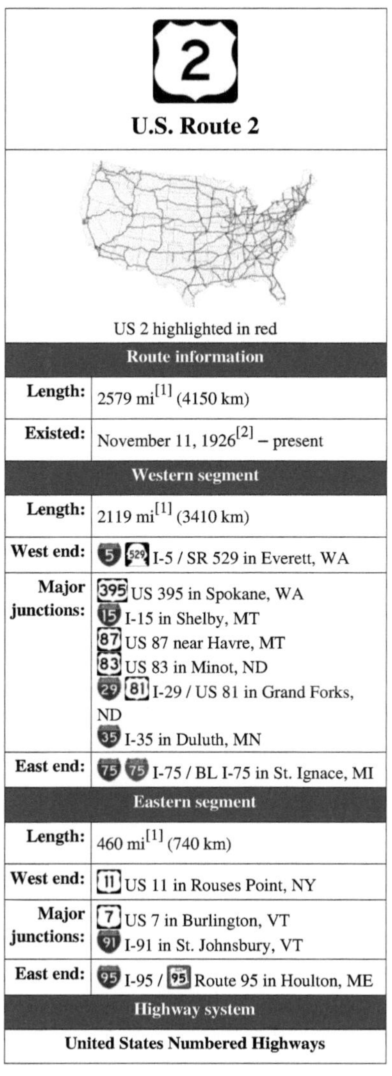

U.S. Route 2	
US 2 highlighted in red	
Route information	
Length:	2579 mi[1] (4150 km)
Existed:	November 11, 1926[2] – present
Western segment	
Length:	2119 mi[1] (3410 km)
West end:	I-5 / SR 529 in Everett, WA
Major junctions:	US 395 in Spokane, WA I-15 in Shelby, MT US 87 near Havre, MT US 83 in Minot, ND I-29 / US 81 in Grand Forks, ND I-35 in Duluth, MN
East end:	I-75 / BL I-75 in St. Ignace, MI
Eastern segment	
Length:	460 mi[1] (740 km)
West end:	US 11 in Rouses Point, NY
Major junctions:	US 7 in Burlington, VT I-91 in St. Johnsbury, VT
East end:	I-95 / Route 95 in Houlton, ME
Highway system	
United States Numbered Highways	

U.S. Route 2 (**US 2**) is an east–west U.S. Highway spanning 2579 miles (4150 km) across the northern continental United States. US 2 consists of two segments connected by various roadways in southern Canada. Unlike some routes, which are disconnected into segments because of encroaching Interstate Highways, the two portions of US 2 were designed to be separate in the original 1926 highway plan.

The western segment of US 2 has its western terminus at State Route 529 (Maple Street) in Everett, Washington and its eastern terminus at Interstate 75 in St. Ignace, Michigan. The eastern segment of US 2 has its western terminus at

US 11 in Rouses Point, New York and its eastern terminus at Interstate 95 (I-95) in Houlton, Maine.

As its number indicates, it is the northernmost east–west U.S. Route in the country. It is the lowest primary-numbered east–west U.S. Route, whose numbers otherwise end in zero, and was so numbered to avoid a US 0.[3]

Sections of US 2 in New England were once **New England Route 15**, part of the New England road marking system.

Route description

Western segment

Lengths

	mi	km
WA	331	533
ID	80	129
MT	664	1068
ND	354	570
MN	264	425
WI	120	193
MI	306	492
Total	2119	3410

The western segment of US 2 extends from the Upper Peninsula of Michigan across the northern tier of the lower 48 states. Most of the western route was built roughly paralleling the Great Northern Railway. US 2 adopted the railway's route nickname "The Highline" as the most northern crossing in the U.S.

The Adventure Cycling Association's Northern Tier Bicycle Route is a bicycle touring route which follows or parallels US 2 for over 600 miles (970 km), most notably a 550 miles (890 km) stretch between Columbia Falls, Montana and Williston, North Dakota.

Washington

Within Washington state, US 2 is the northernmost all-season highway through the Cascade Mountains. It begins at Interstate 5 and State Route 529 in Everett, and travels east via Stevens Pass, Wenatchee, and Spokane to the border in Newport.

Idaho

Shortly after entering Idaho from the east, US 2 Crosses Priest River. US 2 follows Pend Oreille River until it empties into Lake Pend Oreille. US 2 intersects Idaho Route 57 in the town of Priest River at mile 5.8. US 2 intersects US 95 at mile 28.4 in the town of Sandpoint. The two routes are duplexed for 36.2 miles (58.3 km). At Three Mile Corner, US Route 2 continues southeast for 15.8 miles (25.4 km) where it crosses into Montana.

Montana

US 2 is a vital northern corridor for Montana. The road travels through some of the most beautiful land in the state, especially in its western half, and has more of its mileage within Montana than in any other state. It passes through three Indian reservations, comes very close to two others, and skirts the southern border of Glacier National Park. Most of the Montana segment of US 2 runs close to the northern BNSF Railway main line, and parts of the highway show up in the Microsoft Train Simulator depiction of the Marias Pass route.

US 2 passes into Montana 10 miles (16 km) from Troy, a small city. It is also near the lowest point in Montana, where the Kootenai River leaves the state. The first large city the highway comes to is Libby. After this it meanders south and east towards Kalispell, a city of about 20,000 residents north of Flathead Lake, the largest freshwater lake west of the Mississippi River. From there the highway passes through the southern end of Glacier National Park and follows the Middle Fork of the Flathead River. After crossing the continental divide at Marias Pass west of East Glacier, the highway exits the Rocky Mountains and begins its trek through the northern plains. Just before entering East Glacier, it crosses the boundary of the Blackfeet Indian Reservation of northern Montana.

As the highway enters the Great Plains, the first town it encounters is Browning, the largest settlement on the Blackfeet Indian Reservation. From here to the North Dakota border, the surrounding area is also known as "The Hi-Line" to Montanans from the early GN railway route. It next travels through Cut Bank to Shelby, where it becomes the northern border of the area known as the "Golden Triangle" in Montana. This area is one of the most productive farming regions in the country. From Shelby it hits a string of small towns before it goes on to Havre, near the geographical center of the road in the state. Just south of Havre and off the highway about fifteen miles (24 km) is the Rocky Boy Indian Reservation. The highway continues east to Malta, before which it travels through the Fort Belknap Indian Reservation. From Malta, the highway continues on to Glasgow, just north of Fort Peck Dam, and then into the Fort Peck Indian Reservation. The highway stays within the reservation for much of its remaining trip through Montana. On the reservation it goes through Wolf Point and Poplar, and then exits the reservation a short distance before leaving the state. The final town of Bainville says goodbye to the highway as it leaves the state, near the confluence of the Missouri and Yellowstone Rivers.

North Dakota

US 2 is an east–west highway that runs through North Dakota's northern tier of larger cities: Williston, Minot, Devils Lake, and Grand Forks. These cities are about 75 to 100 miles (121 to 160 km) north of North Dakota's southern tier of larger cities located on Interstate 94: Dickinson, Bismarck/Mandan, Jamestown, and Fargo/West Fargo. Each city (or pair) in each tier is separated by about 75 to 125 miles (121 to 201 km). This alignment is probably the reason that two major east–west four-lane highways have developed in North Dakota.

US 2 intersects with two north–south four-lane highways in North Dakota: US 83 [4] at Minot and I-29 at Grand Forks. In addition, it junctions with three other U.S. Highways that, except for shorter stretches that are four-laned, are mostly two-lane highways in North Dakota: US 85 at Williston, US 52 at Minot, US 281 at Churchs Ferry (west of Devils Lake), and US 81 at Grand Forks. All six of these highways provide routes either to the border at Mexico or deep into southern USA.

North Dakota has been converting sections of US 2 from two lanes to four lanes for many years. The section from Grand Forks to Minot was completed several years ago. The section from Minot to Williston was completed in the summer of 2008 in a campaign that began a few years ago and was labeled "Across the State in Two Thousand Eight". Actually, US 2 is four-laned from North Dakota's eastern edge to just past Williston, a stretch of about 343 miles (552 km), leaving the remaining 12 miles (19 km) to the Montana border as a two-lane highway. North Dakota's governor has said that North Dakota will four-lane the remaining stretch if Montana is willing to continue the four-lane project from the border into their state.

Between Williston and Minot, US 2 provides several high points where one can view graceful and beautiful landscape for many miles in all directions. Between Minot and Grand Forks, US 2 provides an ever-changing mix of agricultural farm and pasture land, native wetlands, and small lakes set on a gently rolling landscape. US 2 also passes near a large lake near Devils Lake.

In Rugby, North Dakota, the highway passes the location designated in 1931 as the geographical center of North America. The monument marking the geographic center had to be relocated in 1971 when US 2 was converted from two lanes to four lanes.[5]

Minnesota

The portion of US 2 from Cass Lake to Bemidji is officially designated the Paul Bunyan Expressway. It also intersects US 169 and the Mississippi River in Grand Rapids, Minnesota. At the crossing between Duluth, Minn. and Superior, Wisc., the highway crosses the Richard I. Bong Memorial Bridge, about 8300 feet (2500 m) in length—roughly 11800 feet (3600 m) in length when the above land approaches are included.

Of the 266 miles (428 km) of US 2 in Minnesota, 146 miles (235 km) have four lanes, mostly located in the northwest part of the state.

Legally, the Minnesota section of US 2 is defined as Routes 8 and 203 in Minnesota Statutes §§161.114(2) and 161.115(134).[6] [7]

Wisconsin

After crossing the Bong Bridge and entering into the city of Superior, Wisconsin's western segment of the highway joins Belknap Street. After crossing the midsection of Superior, US 2 merges with US 53 for a few miles following East 2nd Street out of the city. Ten miles outside of Superior, US 53 and US 2 part ways. US 53 veers south toward Eau Claire, while US 2 continues to the city of Ashland and ultimately to the Wisconsin–Michigan state line at the city of Ironwood. An eastern segment of US 2 re-enters Wisconsin 4 miles (6 km) northwest of Florence and proceeds concurrently with US 141 for 14.5 miles (23.3 km) until exiting Wisconsin again near Iron Mountain, Michigan.

Michigan

US 2 enters Michigan at the city of Ironwood and runs east to the town of Crystal Falls, where it turns south and re-enters Wisconsin northwest of Florence. It re-enters Michigan north of Iron Mountain and continues through the Upper Peninsula of Michigan to the cities of Escanaba, Manistique, and St. Ignace. Along the way, it cuts through the Ottawa and Hiawatha National Forests and follows the northern shore of Lake Michigan. It ends at I-75, just north of the Mackinac Bridge in St. Ignace.

Eastern segment

Lengths

	mi	km
NY	0.88	1.42
VT	150.60	242.37
NH	35.43	57.02
ME	273.64	440.38
Total	460.55	741.19

The eastern segment of US 2 traverses the northern New England states.[8]

Vermont

The road starts up at US 11, just one mile (1.6 km) south of the Canadian border at Rouses Point in Champlain New York. From there it crosses Lake Champlain on the Rouses Point Bridge into Grand Isle County, Vermont, traversing the length of the county and crossing the lake over several bridges until it reaches the mainland in Milton and Chittenden County. From there it travels south to Burlington, where it begins to closely parallel Interstate 89 all the way to Montpelier and Washington County. At Montpelier, the road turns north-eastward, crossing into Caledonia County and passing through Saint Johnsbury. It then passes into Essex County, and eventually crosses the Connecticut River from Guildhall, Vermont into Lancaster, New Hampshire.

US 2 in Vermont

New Hampshire

Once into New Hampshire, the road continues southeastward, passing through Jefferson (home to several small amusement parks and roadside attractions, such as Santa's Village and Six Gun City [9]) before heading more easterly, skirting the northern edge of the White Mountain National Forest into Gorham, where it meets Route 16, the major north–south roadway through the eastern half of the forest and past Mount Washington. From Gorham, the road travels east along the southern banks of the Androscoggin River to Shelburne and eventually crossing into Gilead, Maine. Throughout its entire 35-mile (56 km) stretch, the New Hampshire portion of US 2 is exclusively in Coos County.

Maine

US 2 travels from Gilead to Houlton near the Houlton International Airport. US 2 ends just west of the Canadian border at the termini of I-95 and Route 95.

History

A large portion of the western segment of US 2, and a shorter piece of the eastern segment, follows the old Theodore Roosevelt International Highway. This auto trail, named in honor of the late former president and naturalist Theodore Roosevelt, was organized in February 1919 to connect Portland, Maine with Portland, Oregon.[10] The route taken by this highway left Portland, Maine to the northwest, crossing New England via Littleton and Montpelier to Burlington. It crossed Lake Champlain on the Burlington-Port Kent Ferry and headed west across upstate New York, through Watertown and Rochester to Buffalo. After crossing southern Ontario, the highway re-entered the U.S. in Detroit, running northwest and north via Saginaw and Alpena to the Upper Peninsula, where it turned west along the northern tier of the country. This portion took the route past Duluth, Minot, Havre, and Glacier National Park to Spokane. In order to reach Portland, Oregon, the highway turned south in Washington via Walla Walla to Pendleton, where it headed west again via the Columbia River Highway to Portland. The last piece of the highway to be completed was over Marias Pass through Glacier National Park; cars were carried through the park on the Great Northern Railway until 1930.[11] [12] [13]

The first inter-state numbering for the Roosevelt Highway was in New England, where the New England road marking system was established in 1922. Route 18 followed the auto trail from Portland northwest to Montpelier, where it continued to Burlington via Route 14. Many of the states along the route also assigned numbers to the highway; for instance, New York labeled their portion Route 3 in 1924.[12] [14] The Joint Board on Interstate Highways distributed its preliminary plan in 1925, in which a long section of the highway was labeled US 2, from St. Ignace, Michigan west to Bonners Ferry, Idaho. East of St. Ignace, instead of crossing to the Lower Peninsula like the Roosevelt Highway, the proposed Route 2 traveled north to the international border at Sault Ste. Marie. It reappeared at Rouses Point, New York, following Route 30 and then rejoining the auto trail between Burlington and Montpelier. US 2 and the Roosevelt Highway both connected Montpelier to St. Johnsbury, but the latter took a direct path along Route 18, while the former was assigned to Route 25 to Wells River, where it overlapped proposed US 5 north to St. Johnsbury. There, where the Roosevelt Highway turned southeast to Portland, Route 2 continued east along Route 15 to Bangor and Route 1 to Calais, then heading north on Route 24 to end in Houlton.[15]

By the time the U.S. Highway system was finalized in late 1926, one relatively minor change had been made to US 2; it was swapped with US 1 between Bangor and Houlton, Maine, placing US 2 along the entire portion of Route 15 east of St. Johnsbury. Several other major parts of the auto trail received numbers, most notably US 30 from Portland, Oregon east to Pendleton, US 195 in eastern Washington, and US 23 in Michigan's Lower Peninsula.[16] [17] In the mid-1930s, much of New York's portion of the road became US 104, and the part southeast of Littleton, New Hampshire to Portland, Maine became US 302, but by far the longest piece was that followed by US 2 between St. Ignace and Bonners Ferry. In 1946, US 2 was extended west of its original western terminus in Bonners Ferry in Idaho to Everett in Washington via Spokane along what was then Alternate US 10.

Michigan

US 2 was in the original 1925 U.S. Highway Plan by the Bureau of Public Roads[1] and was first commissioned in Michigan in 1926.[18]

US 2 originally ran in Michigan from Ironwood to St. Ignace, the same termini as today. The highway has undergone many realignments, mostly minor, between those cities since 1926. In 1933, the section between St. Ignace and Sault Ste. Marie was relocated along Mackinac Trail.[18]

In 1957, the first segment of a new freeway opened between St. Ignace and Sault Ste. Marie. It ran from Evergreen Shores, north of St. Ignace, to present-day M-123 and replaced the former route on State St. and Mackinac Trail. Over the next six years, US 2 was moved from Mackinac Trail onto the new freeway as new sections opened. Beginning in 1961, the freeway was concurrently signed as an extension of I-75. The freeway was completed in 1963.[18]

The eastern terminus of US 2 in Michigan was truncated back to St. Ignace in 1983, removing it entirely from the I-75 freeway.[18]

Eastern segment

Before being designated as US 2, most of the current alignment was called New England Interstate Route 15' from Danville, Vermont eastward to Maine. The portion of the old Route 15 that did not become part of US 2 was designated as Vermont Route 15.

Other sections of US 2 in Vermont that were not part of New England Route 15 were parts of other former New England Interstate routes: Route 18 between Montpelier and Danville; Route 14 between Burlington and Montpelier; and Route 30 between Alburgh and Burlington.

Major intersections

Western segment

- Interstate 5 in Everett, Washington
- U.S. Route 97 near Cashmere, Washington
- Interstate 90/U.S. Route 395/U.S. Route 195 in Spokane, Washington
- U.S. Route 95 in Sandpoint, Idaho
- U.S. Route 93 in Kalispell, Montana
- U.S. Route 89 in Browning, Montana
- Interstate 15 in Shelby, Montana
- U.S. Route 85 in Williston, North Dakota
- U.S. Route 52 in Minot, North Dakota
- U.S. Route 83 in Minot, North Dakota
- U.S. Route 281 near Churchs Ferry, North Dakota
- U.S. Route 81 in Grand Forks, North Dakota
- Interstate 29 in Grand Forks, North Dakota
- U.S. Route 71 in Bemidji, Minnesota
- Interstate 35 in Duluth, Minnesota
- U.S. Route 51 in Hurley, Wisconsin

Eastern segment

- Interstate 89/U.S. Route 7 in Colchester, Vermont
- Interstate 91/U.S. Route 5 in St. Johnsbury, Vermont
- Interstate 95 in Newport, Maine

See also

- Bannered routes of U.S. Route 2

References

[1] Droz, Robert V. (June 1, 2010). "Sequential List with Termini and Lengths in Miles" (http://www.us-highways.com/us1830.htm). *US Highways From US 1 to US 830.* . Retrieved June 6, 2010.

[2] Weingroff, Richard F. (January 9, 2009). "From Names to Numbers: The Origins of the U.S. Numbered Highway System" (http://wwwcf.fhwa.dot.gov/infrastructure/numbers.cfm). Federal Highway Administration. . Retrieved April 21, 2009.

[3] "Ask the Rambler: What Is The Longest Road in the United States?" (http://www.fhwa.dot.gov/infrastructure/longest.cfm). Federal Highway Administration. December 29, 2008. . Retrieved April 14, 2009.

[4] 2, US 83, 22nd Avenue intersection at Minot (http://maps.google.com/maps?num=100&hl=en&safe=off&client=firefox-a&q=minot,+north+dakota&ie=UTF8&hq=&hnear=Minot,+Ward,+North+Dakota&gl=us&ei=sLd5TKWrCsWinQf-wrmdCw&ved=0CCUQ8gEwAA&ll=48.207158,-101.293795&spn=0.011025,0.033023&z=16lUS)

[5] "Geographical Center of North America" (http://www.rugbynorthdakota.com/AttractionDetail.aspx?AttractionID=6). Rugby, ND: Rugby Area Chamber of Commerce. . Retrieved June 6, 2010.

[6] Minnesota State Legislature (2009). "§ 161.114, Constitutional Trunk Highways" (https://www.revisor.mn.gov/statutes/?id=161.114). *Minnesota Statutes*. Minnesota Office of the Revisor of Statutes. . Retrieved June 6, 2010.

[7] Minnesota State Legislature (2009). "§ 161.115, Additional Trunk Highways" (https://www.revisor.mn.gov/statutes/?id=161.115). *Minnesota Statutes*. Minnesota Office of the Revisor of Statues. . Retrieved June 6, 2010.

[8] Sanderson, Dale (March 7, 2010). "End of US highway 2 (eastern segment)" (http://www.usends.com/00-09/002e/002e.html). *Endpoints of US highways.* . Retrieved June 6, 2010.

[9] http://www.sixguncity.com/

[10] Skidmore, Max J. (2006). *Moose Crossing: Portland to Portland on the Theodore Roosevelt International Highway*. Hamilton Books. ISBN 0761835105.

[11] Clason Map Company (1923). *Midget Map of the Transcontinental Trails of the United States* (http://www.fhwa.dot.gov/infrastructure/midgetmap.htm) (Map). . Retrieved June 6, 2010.

[12] Rand McNally (1926). *Auto Road Atlas* (Map).

[13] Hendrix, Mike; Hendrix, Joyce (July 17, 2007). "Marias Pass straddling the Continental Divide on US 2 in Montana" (http://www. travellogs.us/2007Logs/Montana 2007/123-Marias Pass/123-Marias Pass.htm). . Retrieved June 6, 2010.

[14] "New York's Main Highways Designated by Numbers". *New York Times*: p. XX9. December 21, 1924.

[15] Report of Joint Board on Interstate Highways, October 30, 1925, Approved by the Secretary of Agriculture, November 18, 1925

[16] Bureau of Public Roads (1926) (PDF). *United States System of Highways* (http://www.okladot.state.ok.us/hqdiv/p-r-div/maps/ misc-maps/1926us.pdf) (Map). . Retrieved May 10, 2008.

[17] "United States Numbered Highways". *American Highways* (AASHO). April 1927.

[18] Bessert, Christopher J. (January 31, 2009). "Highways 1 through 9" (http://www.michiganhighways.org/listings/MichHwys01-09. html#US-002W). *Michigan Highways*. . Retrieved June 6, 2010.

External links

- US 2 endpoint photos (http://usends.com/00-09/002e/002e.html)
- Michigan US 2 endpoint photos (http://www.state-ends.com/michigan/us2/)
- Former Michigan US 2 eastern terminus (now I-75 northern terminus) (http://www.state-ends.com/michigan/ i75/)

		Main U.S. Routes																	
	1	2	3	4	5	6	7	8	9	10	11	12	13	14	15	16	17	18	19
20	21	22	23	24	25	26	27	28	29	30	31	32	33	34	35	36	37	38	
40	41	42	43	44	45	46		48	49	50	51	52	53	54	55	56	57	58	59
60	61	62	63	64	65	66	67	68	69	70	71	72	73	74	75	76	77	78	79
80	81	82	83	84	85		87		89	90	91	92	93	94	95	96	97	98	99
101	163						400				412				425				
Lists	U.S. Routes • Bannered • Divided • Bypassed																		

Browse numbered routes

U.S._Route_1_in_Maine

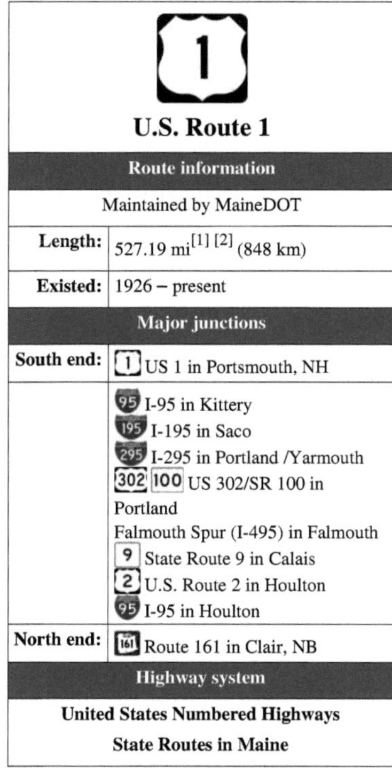

Route information	
Maintained by MaineDOT	
Length:	527.19 mi[1] [2] (848 km)
Existed:	1926 – present
Major junctions	
South end:	US 1 in Portsmouth, NH
	I-95 in Kittery I-195 in Saco I-295 in Portland /Yarmouth US 302/SR 100 in Portland Falmouth Spur (I-495) in Falmouth State Route 9 in Calais U.S. Route 2 in Houlton I-95 in Houlton
North end:	Route 161 in Clair, NB
Highway system	
United States Numbered Highways **State Routes in Maine**	

In the U.S. state of Maine, U.S. Route 1 is a major north–south state highway serving the eastern part of the state. It parallels the Atlantic Ocean from New Hampshire north through Portland, Brunswick, and Belfast to Calais, and then the St. Croix River and the rest of the Canadian border via Houlton to Fort Kent. The portion along the ocean, known as the **Coastal Route**, provides a scenic alternate to Interstate 95.

Route description

Route 1 enters Maine from New Hampshire by bridging the Piscataqua River at Kittery on the Memorial Bridge. However, this bridge was abruptly closed in July, 2011 due to structural defects with a new bridge planned to replace it soon. Now, motorists can only enter Maine by either using the I-95 bridge or the Sarah Mildred Long bridge which carries US 1 Bypass. Following the sandy southern Maine coast, the highway bridges the Cape Neddick River in Cape Neddick, Josias River and Ogunquit River in Ogunquit, the Webhannet River in Wells, and Merriland River, Mousam River and Kennebunk River in Kennebunk. After bridging the Saco River between Biddeford and Saco the highway bridges the Nonesuch River in Scarborough.

In South Portland, US 1 merges with I-295 at exit 4 and continues north through downtown Portland to Tukey's Bridge, now on I-295, before separating at exit 9. The **Charles Loring Highway** is part of US 1 in Portland, Maine. Like Loring Air Force Base it is named for Charles J. Loring, Jr..

North of Portland, the highway bridges the Presumpscot River in Falmouth, the Royal River in Yarmouth, and the Cousins River in Freeport before following the Androscoggin River through Brunswick and bridging the Kennebec

River between Bath and Woolwich. The section between Brunswick (at its junction with US 201) and Bath is a four lane freeway, and the route continues as a four lane expressway through most of Bath, then crosses on a two lane viaduct before becoming four lanes again on the bridge over the Kennebec River. It then becomes a three lane road (with center turning lane) through Woolwich and then reverts back to two lanes after that as it continues north towards Wiscasset. The highway bridges the Sheepscot River in Wiscasset, the Damariscotta River in Damariscotta, the Medomak River in Waldoboro, and the Saint George River in Thomaston before reaching Rockland. In Rockland there is a bypass of downtown (US 1A) which travels along Broadway and Maverick Streets, while US 1 itself has a one way pair with Main Street and Union Street in the downtown Rockland area (Main Street traffic goes north with two lanes, while Union St. traffic is southbound). The highway follows the coast of Penobscot Bay bridging the Goose River in Rockport, the Ducktrap River in Lincolnville, the Little River in Northport, and the Passagassawakeag River in Belfast before bridging the Penobscot River at Bucksport. The Atlantic coast is less frequently visible as the highway bridges the Orland River in Orland, the Union River in Ellsworth, Sullivan Harbor, the Narraguagus River in Cherryfield, the Harrington River in Harrington, the Pleasant River at Columbia Falls, the Indian River in Addison, and the Chandler River in Jonesboro.

After bridging the East Machias River and Machias River in Machias the highway turns inland along Passamaquoddy Bay to bridge the Orange River in Whiting, the Dennys River in Dennysville and the Pennamaquan River in Pembroke. The highway then follows the Saint Croix River through Calais and turns inland at Woodland. The interior route bridges Grand Falls Flowage at Princeton, the Meduxnekeag River in Houlton, the North Branch Meduxnekeag River in Monticello, and the Aroostook River in Presque Isle before following the Saint John River upstream from Van Buren to Fort Kent.

The monument marking the northern terminus in Fort Kent, ME

History

US 1 south of Calais was initially part of the Atlantic Highway, and became Route 1 when the New England road marking system was established in 1922.[3] The northward continuation from Calais was later designated as part of **Route 24**. In the original plan, Route 24 was to run from Brunswick to Moosehead Lake in Greenville. By 1925, however, Maine had transferred the Route 24 designation to a completely new alignment on the eastern edge of the state, running from Calais to Madawaska at a border crossing with Edmundston, New Brunswick.

The initial 1925 plan for the U.S. Highway system took US 1 along the better-quality inland route (then Route 15)[4] between Bangor and Houlton, and placed US 2 on the coastal route.[5] This changed in the final 1926 plan, when the inland shortcut — now generally followed by Interstate 95 - became part of US 2.[6]

The Waldo-Hancock Bridge opened in 1931,[7] allowing US 1 to bypass Bangor; the old route became US 1A.

The portion between Portland and Brunswick was rebuilt, mainly as a four-lane divided highway, in the 1950s, and later absorbed into I-95 (now I-295). A freeway from Brunswick east to Bath was built in the 1960s.

References

[1] Maine State Route Log via floodgap.com (http://www.floodgap.com/roadgap/me/r?1)

[2] American Association of State Highway and Transportation Officials, United States Numbered Highways (http://cms.transportation.org/?siteid=68&pageid=1760), 1989 Edition

[3] New York Times, Motor Sign Uniformity, April 16, 1922, p. 98

[4] Rand McNally Auto Road Atlas, 1926, accessed via the Broer Map Library (http://www.broermapsonline.org/members/)

[5] Report of Joint Board on Interstate Highways, October 30, 1925, Approved by the Secretary of Agriculture, November 18, 1925

[6] United States System of Highways, November 11, 1926

[7] Maine Department of Transportation, Waldo-Hancock Bridge (http://www.maine.gov/mdot/covered-bridges/waldo.php), accessed October 2007

U.S._Route_202

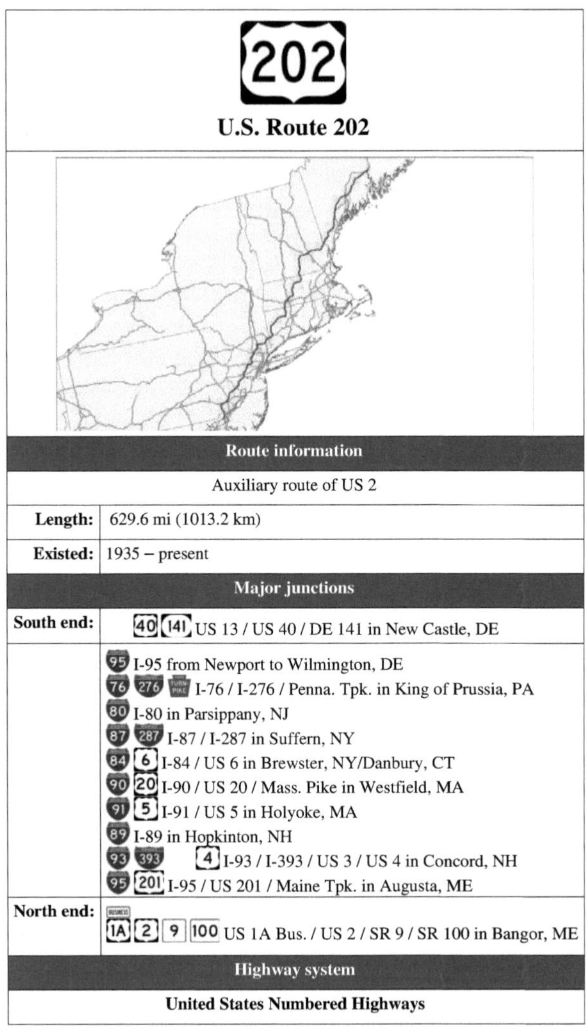

Route information	
Auxiliary route of US 2	
Length:	629.6 mi (1013.2 km)
Existed:	1935 – present
Major junctions	
South end:	US 13 / US 40 / DE 141 in New Castle, DE
	I-95 from Newport to Wilmington, DE I-76 / I-276 / Penna. Tpk. in King of Prussia, PA I-80 in Parsippany, NJ I-87 / I-287 in Suffern, NY I-84 / US 6 in Brewster, NY/Danbury, CT I-90 / US 20 / Mass. Pike in Westfield, MA I-91 / US 5 in Holyoke, MA I-89 in Hopkinton, NH I-93 / I-393 / US 3 / US 4 in Concord, NH I-95 / US 201 / Maine Tpk. in Augusta, ME
North end:	US 1A Bus. / US 2 / SR 9 / SR 100 in Bangor, ME
Highway system	
United States Numbered Highways	

U.S. Route 202 is a highway stretching from Delaware to Maine, also passing through the states of Pennsylvania, New Jersey, New York, Connecticut, Massachusetts, and New Hampshire.

The road has borne the number 202 since at least 1935. Before this, sections of the road were designated U.S. Route 122, as it intersected U.S. Route 22. Its current designation is based on its intersection with US 2 in Maine.

This route is considerably longer than the eastern segment of US 2, making it one of several 3-digit US routes to be longer than their parent routes.

Route description

Delaware

US 202 begins at an interchange with US 13/US 40 south of Wilmington. It runs north along the same road as Delaware Route 141, then joins with Interstate 95 through Wilmington. North of the city, it exits the freeway onto Concord Pike, heading north; Delaware Route 202 also continues south from this point. US 202 continues north towards the state line as a six-lane arterial road and is lined with numerous strip malls and "big-box stores".

Pennsylvania

US Route 202, traveling south near Chesterbrook, Pennsylvania.

US 202 continues north toward West Chester, joining with US 322 north of U.S. Route 1. South of West Chester, US 202/322 exits onto a limited-access bypass of the borough; that is the old West Chester By-Pass, and includes a grade-level intersection at Matlack Street. North of West Chester, US 322 exits, and US 202 continues north as a freeway towards Frazer, where it interchanges with U.S. Route 30 and bends east to head towards Malvern and King of Prussia. The stretch between Valley Road in Paoli and King of Prussia was recently widened to three lanes in each direction. In King of Prussia, the highway forms a large, complicated interchange with the Schuylkill Expressway, the Pennsylvania Turnpike, and U.S. Route 422.

The freeway then transitions into a divided highway, passing the King of Prussia Mall and heading northeast through commercial areas before splitting into a one-way pair through the streets of Bridgeport and Norristown, crossing the Schuylkill River in the process.

US Route 202 (perpendicular), at the expressway terminus in King of Prussia. Twelve lanes move traffic through the intersection with Gulph Road.

North of Norristown, US 202 continues as a two-lane road heading northeast through the Philadelphia suburbs, passing through Blue Bell and Lower Gwynedd, where it becomes a four-lane full-access highway for about two miles (3 km). East of Lansdale, in Montgomeryville, it briefly joins with Pennsylvania Route 309, splitting off at a large, complex intersection locally known as "Five Points". It continues northeast towards Doylestown, bypassing the town first on a wrong-way concurrency on PA 611 then on a short bypass that ends at PA 313. It continues as a two-lane road to New Hope, crossing the Delaware River on the New Hope-Lambertville Toll Bridge.

Construction began in November 2008 on a parkway project between PA 63 in Lower Gwynedd and the existing cloverleaf interchange at the **US 202 Bypass** and PA 611 near Doylestown. This parkway will consist of a four-lane stretch between PA 63 and PA 463 and a two-lane parkway the rest of the way that bypasses the boroughs of Chalfont and New Britain.

New Jersey

On the toll bridge, US 202 has two lanes in each direction. It continues a northeasterly course for about 5.7 miles (9.2 km) as a freeway. This segment of US 202 was earlier called the **202 bypass** (as it bypassed the New Hope-Lambertville area) from its original route. The old section of 202 between New Hope and Ringoes, New Jersey is now NJ 179 which is also Old York Road, the first roadway to connect New York City and Philadelphia, Pennsylvania. In 1953, this section of Old York Road was renumbered US 202. A small section of the US 202 bypass was built in 1965 and the old route was renamed NJ 179. When the western section of the "bypass" was built to the Delaware River, the whole former segment was renamed 179. The section of the new US 202 freeway section ends once it begins to run concurrently with NJ 31 in East Amwell Township. The concurrency runs for five miles (8 km), to Flemington. This stretch, and the 13 miles (21 km) between Flemington and Somerville, is a four-lane divided roadway.

US 202 just inside New Jersey at the NY/NJ state line.

At Somerville, the road merges with US 206 at a now-reconfigured Somerville Circle. Parts of the old traffic circle, which also carries NJ 28, remain below the US 202 flyover. US 202 splits northeastward from US 206 at Bedminster Township and again becomes a two-lane road.

From here to the state line, US 202 parallels, and has largely been supplanted by, I-287, which during its construction dumped traffic onto US 202. US 202 continues through Morristown to Morris Plains with an intersection with NJ 53. Interestingly, with a few exceptions, US 202 is maintained by counties rather than the New Jersey Department of Transportation north of NJ 53.

The following sections are state-maintained:

- At the I-80 interchange
- At the US 46 intersection
- Along the NJ 23 concurrency
- At the I-287 interchange in Oakland

US 202 continues past Boonton along the Boonton Turnpike to Wayne, where it then picks up NJ 23 for about two miles (3 km) and then exits on Black Oak Ridge Road. It then follows the Paterson-Hamburg Turnpike and Ramapo Valley Road (more or less paralleling the Ramapo River through Oakland) to Mahwah before crossing the New York state line on the Franklin Turnpike.

New York

US 202 is mostly designated east–west in New York, owing to its greater coverage in those directions.

Franklin Turnpike becomes Orange Avenue in Suffern, and US 202 continues to a block-long wrong-way concurrency with NY 59 before tailing off on Wayne Avenue and heading east toward Haverstraw. Most of this stretch is two-lane road.

At Haverstraw, US 202 turns north along US 9W to Bear Mountain and then crosses the Bear Mountain Bridge, running concurrently with US 6, the Grand Army Highway. The two wind around Anthony's Nose, briefly forming New York's only three-way

View of the Bear Mountain Bridge from Bear Mountain

concurrency of U.S. highways with US 9 at Peekskill. Afterwards, the two separate for several miles, with US 202 taking the more southerly route through Somers. The highways reunite at Brewster and become a four-lane road for their last few miles before the state line, taking in NY 121 in the process.[1]

Connecticut

At Danbury, US 6 and 202 climb up onto I-84, which had just been joined by the north–south US 7, making a four-way concurrency. Until recently, 202 ran with U.S. Route 7, however recent construction has separated the two highways. 202 splits from I-84 and US 6 at Exit 2. It is a two-lane road in southern Brookfield as it follows Federal Road. The US 7 freeway continues for another 6.5 miles (10.5 km) before it rejoins US 202 at the Brookfield/New Milford town line. The now rejoined US 7 and 202 approach New Milford in bucolic Litchfield County, where they once again split.

US 202 continues through Torrington and on to Cherry Brook, where it then runs concurrently with US 44 for several miles before turning northward at Avon. For the run to the state line, US 202 runs concurrently with Route 10.

Massachusetts

Unlike elsewhere in New England, US 202 is posted as a south–north highway in Massachusetts, as the highway runs mostly in those directions for its length through the state.

US 202 and Route 10 enter the Bay State at the "Congamond Notch" in Southwick, a southward jog in the state line that includes Congamond Lake. North of Westfield, US 202 turns eastward toward Holyoke and Belchertown. It then heads north along the west side of the Quabbin Reservoir through New Salem toward Athol. This section of US 202 has been dubbed the Daniel Shays Highway, named for a Revolutionary War veteran who led an insurrection against the state government of Massachusetts. US 202 meets Route 2 at Orange, and runs along the two-lane freeway to Phillipston. There, it diverges to the north again as a two-lane road.

In Massachusetts, US 202 passes through the municipalities of Southwick, Westfield, Holyoke, South Hadley, Granby, Belchertown, Pelham, Shutesbury, New Salem, Orange, Athol, Phillipston, Templeton, and Winchendon.

New Hampshire

US 202 is posted as an east–west highway in New Hampshire. It remains a two-lane highway for most of its length in the Granite State.

It heads north, through Jaffrey, to Hillsborough, where it turns eastward along a concurrency with New Hampshire Route 9. The span of the road between Hillsborough and Hopkinton, which passes through Henniker, is among the most deadly sections of roadway in the state.[2] At Concord, New Hampshire, the state capital, US 202 heads north and picks up a concurrency with US 3 for a short time, and then turns eastward again along Interstate 393, a freeway spur that also carries US 4. The freeway ends short of Chichester, and NH 9 rejoins the two-lane concurrency along with US 4 and 202.

At Northwood, US 202 and NH 9 leave US 4. NH 9 splits off a few miles later, leaving US 202 to continue alone toward Rochester, where the road jumps up onto the Spaulding Turnpike (NH 16) for a short, non-tolled distance. US 202 leaves the turnpike two miles (3 km) before the state line at East Rochester.

Maine

Connected farm in Windham, Maine typical of older residences adjacent to route 202 through rural New England.

US 202 is posted as an east–west highway in Maine.

The highway enters the state by crossing the Salmon Falls River at South Lebanon and bridges the Mousam River in Sanford. The highway then passes through Alfred, Waterboro and Hollis before crossing the Saco River at Salmon Falls. The highway passes through Gorham and crosses the Presumpscot River into South Windham. There is a rotary with U.S. Route 302 at Foster's Corner and an interchange with I-95 at Gray. The highway parallels I-95 through Upper Gloucester to Auburn and crosses the Androscoggin River into Lewiston, passing near the campus of Bates College.[3] A very short stretch through the latter two cities is four-lane highway, but most of its length in the Pine Tree State consists of two-lane road. Its final miles west of Hampden, including the short overlay on I-395, and concurrency with US 1A also include four-laned segments.

The highway passes through Greene, Monmouth, and Winthrop concurrently with State Route 11 and State Route 100, and becomes concurrent with State Route 17 at Manchester. US 202 runs concurrently with U.S. Route 201 as it crosses the Kennebec River at Augusta, and shortly thereafter it picks up State Route 3 and State Route 9. SR 3 splits off at South China, but SR 9 stays with US 202 through Albion, Unity, Troy, Dixmont, Newburgh, and Hampden almost all the way to its terminus in Bangor.[3] When SR 9 turns right onto Summer Street in Bangor with US 1A, US 202 runs 4 more blocks to its eastern terminus at the US 2 rotary at the corner of Hammond, Main, Central and State Streets.

A proposed extension eastward of US 202 would run up State Street in Bangor with US 2, turning east onto Broadway/Oak St across the bridge to Brewer (the unposted US 1A Business route), and then following SR 9 again and then with US 1 to Calais utilizing the under-construction/proposed 'third Calais crossing'. This is part of the Maine East–West Corridor.

History

U.S. Route 122 was created in 1926, connecting US 22 at Haverstraw, New York with Wilmington, Delaware. It became part of US 202 in 1934.

The Piedmont Expressway

In Pennsylvania in the early 1960s, a four-lane expressway was proposed that would follow the US 202 corridor. The "Piedmont Expressway" was to be 59 miles (95 km) long, and would cost approximately $146 million. It was to serve as an outer beltway around the Philadelphia area, similar to the Capital Beltway that encircles Washington, D.C.

See also

Related routes

- U.S. Route 2
- U.S. Route 302

References

[1] DeLorme Mapping Company (1993). *New York State Atlas and Gazetteer*. DeLorme Mapping Company. ISBN 0-89933-230-7.

[2] http://www.cmonitor.com/apps/pbcs.dll/article?AID=/20050605/REPOSITORY/506050432/1031

[3] DeLorme Mapping Company (1988). *The Maine Atlas and Gazetteer*. DeLorme Mapping Company. ISBN 0-89933-035-5.

External links

- New Jersey section of U.S. Route 202 (http://www.state.nj.us/transportation/refdata/sldiag/00000202__-.pdf) Straight Line Diagram from the New Jersey Department of Transportation
- An enlarged view of road jurisdiction of US 202, NJ 124 and CR 510 in Morristown (http://www.state.nj.us/transportation/refdata/sldiag/enlarged_view_10.pdf)
- New Jersey Roads: U.S. Route 202 (http://www.alpsroads.net/roads/nj/log/10.html#202)
- Speed Limits for New Jersey State Roads: U.S. Route 202 in New Jersey (http://www.state.nj.us/transportation/refdata/traffic_orders/speed/rt202.shtm)

> **Browse numbered routes**

Interstate_95_in_Maine

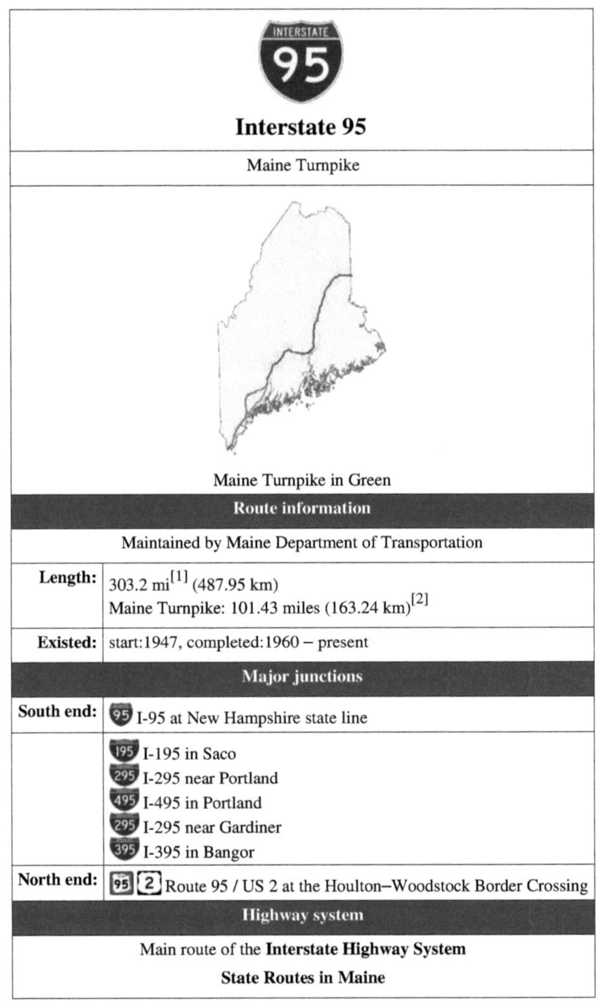

Interstate 95

Maine Turnpike

Maine Turnpike in Green

Route information	
Maintained by Maine Department of Transportation	
Length:	303.2 mi[1] (487.95 km) Maine Turnpike: 101.43 miles (163.24 km)[2]
Existed:	start:1947, completed:1960 − present
Major junctions	
South end:	I-95 at New Hampshire state line
	I-195 in Saco I-295 near Portland I-495 in Portland I-295 near Gardiner I-395 in Bangor
North end:	Route 95 / US 2 at the Houlton−Woodstock Border Crossing
Highway system	
Main route of the **Interstate Highway System** **State Routes in Maine**	

In the U.S. state of Maine, **Interstate 95 (I-95)** is a 305-mile (491 km) long highway running from the New Hampshire border near Kittery, to the Canadian border near Houlton. It is the only two-digit Interstate Highway in Maine. In 2004, the highway's route between Portland and Gardiner was changed so that it encompasses the entire **Maine Turnpike**, which runs from Kittery to Augusta.

Route description

I-95 enters Maine from New Hampshire on the Piscataqua River Bridge, which connects Portsmouth, New Hampshire with Kittery. At York, the highway becomes the Maine Turnpike. The highway runs in a general northeasterly direction, parallel with U.S. 1, at this point. I-95 bypasses the Biddeford/Saco area, with a spur route (Interstate 195) connecting to Old Orchard Beach.

At Scarborough, Interstate 95 meets Interstate 295. The highway turns north, serving the Portland International Jetport and bypassing Portland to the west. At Falmouth, the highway meets unsigned Interstate 495, also called the Falmouth Spur. Until January 2004, I-95 followed the Falmouth Spur and I-295 between Falmouth and Gardiner.

The highway continues north along the Maine Turnpike through Gray to Auburn and Lewiston, which the Turnpike bypasses to the south. The highway then runs in an easterly direction to meet Interstate 295 at Gardiner. From there, I-95 parallels the Kennebec River past Augusta and Waterville. The highway crosses the river at Fairfield and then turns northeast along the Sebasticook River past Pittsfield to Newport.

I-95 then continues east alongside U.S. Route 2 from Newport to Bangor, where Interstate 395 connects to the city of Brewer. The highway runs along the northern edge of Bangor's center, then turns northeast, following the Penobscot River past Orono and Old Town (Prior to the early 1980s, I-95 was a super two highway north of Old Town).

Northbound in Kittery, Maine

The highway continues north, still running near the river, towards Howland. Near Lincoln, Interstate 95 runs north through uninhabited forest land, crossing the Penobscot River at Medway. The highway goes northeast and east, passing a series of small Aroostook County farming towns before reaching Houlton, where it connects to New Brunswick Route 95 and U.S. Route 2 at the international border. North of Bangor, traffic levels drop noticeably, with AADT averaging only about 5,000 in northern Penobscot County and going down to as low as 2,000-4,000 in Houlton.[3]

A small, disused cemetery lies on the road shoulder near Kennebunk. Although it is less than five feet from the roadside, crews have taken care to preserve it, including erecting a fence around the tombstones so that snowplows do not cause any damage.

Speed limits

The Maine Turnpike had a posted speed limit of 70 mph in the early 1970s, but as Maine then had no law against traveling less than 10 mph over the posted speed, the de facto speed limit was 79 mph. In 1974, as part of a federal mandate, the speed limit was reduced to 55 mph, with a new law including a 'less than 10 over' violation. In 1987, Congress allowed states to post 65 mph on rural interstate highways. Following the relaxation, Maine increased its speed limit. In May 2011, a bill was introduced to raise the speed limit from Old Town to Houlton from 65 to 75 mph. It passed, with Maine the first state east of the Mississippi River since the 1970s to establish a 75 mph speed limit.[4] [5]

Exit list

Note: toll rates listed in this exit list are for Class 1 vehicles paying cash. They do not reflect lower rates for drivers using EZ-Pass tags or higher rates for other vehicle classes.

County	Location	Exit #	Destinations	Notes	
			95 continues into New Hampshire southbound		
York	Kittery	1	**103** To SR 103 / Dennett Road - Portsmouth Naval Shipyard, Eliot	Northbound exit and southbound entrance	
		2	**1** **236** **1** US 1 south / SR 236 south / US 1 Byp. south / Traffic Circle – Downtown Kittery		
		3	**1** **236** US 1 north (Coastal Route) / SR 236 north – Kittery, South Berwick	Southbound exit is via exit 2	
		7	**91** **1** To SR 91 to US 1 – The Yorks, Ogunquit, The Berwicks		
			Maine Turnpike York Toll Barrier - Cars $2.00		
		19	**9** **109** SR 9 / SR 109 – Wells, Sanford	Entrance toll $1.00 (northbound only)	
	Kennebunk	25	**35** SR 35 – Kennebunk, Kennebunkport	Entrance toll $1.00	
	Biddeford	32	**111** SR 111 – Biddeford	Entrance toll $1.00	
	Saco	36	**195** I-195 east – Saco, Old Orchard Beach	Entrance toll $1.00	
Cumberland		42	**1** To US 1 – Scarborough	Entrance toll $1.00	
		44	**295** I-295 north – South Portland, Downtown Portland	Northbound exit and southbound entrance; toll $1.00 both directions	
	South Portland	45	**295** **1** **114** To I-295 / US 1 / SR 114 / Maine Mall Road, Payne Road	Entrance toll $1.00	
	Portland	46	**22** **9** To SR 22 / SR 9 (Congress Street) – Portland International Jetport	Entrance toll $1.00	
		47	**25** To SR 25 / Rand Road, Westbrook Arterial	Entrance toll $1.00	
		48	**25** **302** To SR 25 / US 302 / Riverside Street, Larrabee Road	Entrance toll $1.00	
		52	**295** **1** To I-295 / US 1 – Falmouth, Freeport	Toll $1.00 both directions on I-495	
		53	**26** **100** SR 26 / SR 100 – West Falmouth	Entrance toll $1.00	
		63	**202** **115** **4** **26** US 202 / SR 115 / SR 4 to SR 26 – Gray, New Gloucester	Entrance toll $1.00 (southbound only)	
			New Gloucester Toll Barrier - Cars $1.75		
Androscoggin	Auburn	75	**202** **4** **100** US 202 / SR 4 / SR 100 – Auburn		
	Lewiston	80	**196** To SR 196 – Lewiston		
		86	**9** SR 9 – Sabattus, Lisbon		

County		Exit	Destinations	Notes	
Kennebec	**West Gardiner Toll Barrier - Cars $1.25**				
		102	9 126 295 SR 9 / SR 126 to I-295 south – Gardiner, Litchfield	Northbound exit and southbound entrance	
		103	295 9 126 I-295 south to SR 9 / SR 126 – Gardiner, Brunswick	Southbound exit and northbound entrance; toll $1.00 in both directions	
	Augusta	109	202 11 17 100 US 202 / SR 11 / SR 17 / SR 100 – Augusta, Winthrop	Signed as exits 109A (west) and 109B (east) southbound	
		112	8 11 27 SR 8 / SR 11 / SR 27 – Augusta, Belgrade	Signed as exits 112A (south) and 112B (north) northbound	
		113	3 SR 3 – Augusta, Belfast		
		120	Lyons Road - Sidney		
	Waterville	127	11 137 SR 11 / SR 137 – Waterville, Oakland		
		130	104 SR 104 (Main Street) – Waterville, Winslow		
Somerset	Fairfield	132	139 SR 139 – Fairfield, Benton		
		133	201 US 201 – Fairfield, Skowhegan		
Kennebec		138	Hinckley Road - Clinton, Burnham		
Somerset	Pittsfield	150	Somerset Avenue - Pittsfield, Hartland, Burnham		
		157	11 100 7 2 SR 11 / SR 100 to SR 7 / US 2 – Newport, Dexter, Skowhegan		

County	Location	Exit	Destinations	Destinations	Notes
Penobscot		159	Ridge Road - Newport, Plymouth		Southbound exit and northbound entrance
		161	[7] SR 7 – East Newport, Plymouth		
		167	[69] [143] SR 69 / SR 143 – Etna, Dixmont		
		174	[69] SR 69 – Carmel, Winterport		
		180	Cold Brook Road - Hermon, Hampden		
	Bangor	182A	[395] [15] [1A] [9] I-395 / SR 15 south to US 1A / SR 9 – Bangor, Brewer		South end of SR 15 overlap
		182B		[2] [100] To US 2 west / SR 100 west – Hermon	
		183		[2] [100] US 2 / SR 100 (Hammond Street) – Bangor International Airport	
		184		[222] SR 222 (Union Street) / Ohio Street – Bangor International Airport	
		185		[15] SR 15 north (Broadway) – Brewer, Bangor	North end of SR 15 overlap
		186		Stillwater Avenue	No northbound entrance
		187		Hogan Road - Bangor, Veazie	
	Orono	191	Kelly Road - Orono, Veazie		
		193	Stillwater Avenue - Stillwater, Old Town, Orono		
	Old Town	197	[43] SR 43 – Old Town, Hudson		
		199	[16] SR 16 – Alton, Lagrange, Milo		Northbound exit and southbound entrance
		217	[6] [155] SR 6 / SR 155 – Howland, Lagrange		
	Lincoln	227	[2] [6] [116] To US 2 / SR 6 / SR 116 – Lincoln, Mattawamkeag		
	Medway	244	[157] SR 157 – Medway, Millinocket, Mattawamkeag		
	Benedicta	259			Northbound exit and southbound entrance / I 95 in Aroostook county for 1/2 mile
Aroostook	Sherman	264	[158] [11] SR 158 to SR 11 – Sherman, Patten		
	Island Falls	276	[159] SR 159 – Island Falls, Patten		
	Oakfield	286	Oakfield Road - Oakfield, Smyrna Mills		
	Smyrna	291	[2] US 2 – Smyrna		
	Houlton	302	[1] US 1 – Houlton, Presque Isle		
		305	[2] US 2 west – Houlton International Airport, Industrial Park		East end of US 2
			[95] Route 95 east – Houlton–Woodstock Border Crossing, Woodstock		Continuation into New Brunswick, Canada

Maine Turnpike

Tolls

The segment of Interstate 95 from Kittery to Augusta runs along the Maine Turnpike. It is a toll road for all of that length except for a short section in Kittery and York. Flat-fee tolls are paid upon entering the turnpike and there are also barrier tolls in York, New Gloucester, and West Gardiner. The turnpike joined the E-ZPass electronic toll collection network in 2005, replacing the former Maine-only system designated Transpass that was implemented in 1997.[6]

Service areas

There are six service areas on the turnpike, three in each direction. All are open 24 hours and provide food and fuel services. They also have ATMs. Some have small gift shops. The plazas are at the following locations:

- Kennebunk plazas — Northbound and southbound at MP 25 - Food, fuel, gift shop. Original 1972 plazas were replaced during the winter of 2006-2007. Both service plazas open with "food court layout featuring Starbucks coffee, Burger King, Hershey's Ice Cream, a Z-Market convenience store and a Popeye's Chicken on the northbound side and Sbarros Pizza on the southbound side." [7]
- Gray plaza (NB)/Cumberland plaza (SB) - Northbound and Southbound at MP 58 - Food and fuel. Both plazas have been replaced with new service plazas with a Starbucks and a Z-Market convenience store.
- Gardiner plaza — At the I-95/I-295/ME 126 intersection,accessible by both directions of I-95 and I-295. Food court, fuel, gift shop, information.

There is a Rest Area / Tourist Welcome Center located on the turnpike Northbound at MP 3 in Kittery.

There are Weigh Stations located on the turnpike Northbound and Southbound in York at MP 4 (SB) and MP 6 (NB).

There are ramps to/from the northbound turnpike to the Saco Holiday Inn Express Hotel and Conference Center in Saco at MP 35 (Old MP 33 before the southern extension). The ramps are from the original exit 5 which was replaced when I-195 was opened just to the north. The hotel was built on the site of the old toll plaza. Ramps connecting the hotel to/from the southbound turnpike were removed as part of the widening project in the early 2000s, when hotel ownership opted not to pay nearly $1 million to build a new bridge.

Previous to the Gardiner rest area opening, there were rest areas located in Lewiston (Southbound at MP 83) and Litchfield (Northbound at MP 98).

References

[1] Maine State Route Log (http://www.floodgap.com/roadgap/me/r?95) (via floodgap.com)

[2] "Welcome to the Maine Turnpike Authority" (http://www.maineturnpike.com). Maineturnpike.com. . Retrieved 2011-09-19.

[3] "Interstate 95 Annual Average Daily Traffic" (http://www.interstate-guide.com/i-095_aadt.html). Interstate-Guide. . Retrieved 2011-09-19.

[4] Miller, Kevin (May 12, 2011). "Bill would boost speed limit to 75 mph on northern highway" (http://new.bangordailynews.com/2011/05/12/politics/bill-would-boost-speed-limit-to-75-mph-on-northern-highway/). *Bangor Daily News*. . Retrieved May 22, 2011.

[5] Lawmakers OK 75-mph speed limit between Old Town, Houlton, Bangor Daily News, June 29, 2011 (http://bangordailynews.com/2011/06/28/politics/lawmakers-ok-75-mph-speed-limit-between-old-town-houlton/?ref=mostReadBox)

[6] "E-ZPass Information Frequently Asked Questions" (https://ezpassmaineturnpike.com/EZPass/info/faqs.html). .

[7] Turnpike Press Release (http://www.maineturnpike.com/html/news/press_release.html?recordid=129)

External links

- Maine Turnpike Official Site (http://www.maineturnpike.com)
- Steve Anderson's BostonRoads.com: Maine Turnpike (I-95) (http://www.bostonroads.com/roads/me-turnpike)

Sunday_River_(ski_resort)

Sunday River Ski Resort	
Sunday River.	
Location	Newry, Maine, United States
Nearest city	Auburn, Maine
Coordinates	44°28′10″N 70°51′40″W
Top elevation	3140 ft (960 m)
Base elevation	800 ft (240 m)
Skiable area	660 acres (270 ha)
Runs	133
Longest run	3 mi (4.8 km)
Lift system	16 (14 chairs): • 1 Chondola • 4 High-Speed Quads • 5 Quads • 3 Triples • 1 Double • 2 Surface Lifts
Snowfall	155 inches (390 cm) per year
Web site	www.sundayriver.com [1]

Sunday River is a ski resort located in Newry, Maine, in the United States. It is one of Maine's largest and most visited ski resorts. Its vertical drop of 2340 feet (710 m) is the second largest in Maine (after Sugarloaf) and the sixth largest in New England. The resort features 133 trails across eight interconnected mountain peaks, and is serviced by a network of 16 lifts.

Sunday River and its sister resort Sugarloaf/USA have been operated by Boyne Resorts since being sold by American Skiing Company in 2007 for a combined $77 million.[2]

Statistics

- **Important facts**[3]
  - Year Opened: 1959
  - Number of Lifts: 16
  - Chondola (6 person chairs & 8 person gondolas): 1
  - High-speed quads: 4
  - Quad chairs: 5
  - Triple chairs: 3
  - Double chairs: 1
  - Surface lifts: 2
  - Lift capacity: 32,000/hr.
  - Slopeside Hotels: 2 (Jordan Grand Hotel, Grand Summit Hotel).
- **Elevation**
  - Top: 3140 ft (960 m)
  - Bottom: 800 ft (240 m)
  - Vertical drop: 2340 ft (710 m)
  - Longest run: 3.50 miles (5.63 km)
  - Skiable area: 667.7 acres (2.7 km^2)
  - Snowmaking: 92%

- **Types of runs**
  - Beginner: 33%
  - Intermediate: 36%
  - Advanced: 19%
  - Expert: 12%

Terrain

The following describes the different mountains of Sunday River:

- **White Cap**: Located at the eastern end of the resort, it contains mostly difficult terrain consisting of steeps and glades. *White Heat* is an expert run, one of the steepest at Sunday River. *Obsession* is a long Advanced to intermediate trail full of rolls and variation in pitch.

- **Locke Mountain**: The resort's original peak. Features predominantly intermediate trails like *Cascades*, *Monday Mourning* and *Sunday Punch*. The lower section of Locke is also home to a half-pipe and the Rocking Chair terrain park.

- **Barker Mountain**: Features a high-speed quad serving a variety of terrain. *Three Mile Trail* is a beginner run that winds across the mountain. *Ecstasy* and *Right Stuff* are other important trails. Monday Mourning and the race arena parallel it host a variety of high school- and middle school-level alpine ski races. Also services a newschool terrain park on Rocking Chair for skiers and snowboarders featuring a variety of jump and rail elements.

- **Spruce Peak**: Contains two intermediate trails, *Risky Business* and *American Express*. *Downdraft* is a short and steep difficult run that often contains moguls, or is iced over. The peak also offers quick access to the western side of the resort with the beginner trail *Sirius*.

- **North Peak**: Has the mid-mountain Peak Lodge that offers food and beverages. Also contains a high-speed quad and "Chondola" lift that services beginner and intermediate terrain. North Peak is oriented toward "family" skiing. The *Dream maker* trail is a long and wide beginner to intermediate trail that is a favorite for families.

- **Aurora Peak**: A mix of intermediate and difficult trails serviced by a quad chairlift. There is also a double chairlift that connects Aurora to the base of Jordan Bowl.

- **OZ**: A unique bowl with difficult and expert terrain. Runs are a mix of glades and open trails that emphasize natural terrain features and steep pitches.

Sunday River from the Jordan Bowl. Old Speck Mountain on center horizon.

Looking up toward the Jordan Bowl summit.

The top of "Kansas."

- **Jordan Bowl**: Contains a mix of terrain that covers all ability levels. The Jordan Grand Resort Hotel is located off the beginner trail *Lollapalooza*. Other runs include *Rogue Angel* and *Excalibur*.
- **South Ridge**: South Ridge services beginner trails. The South Ridge Ski Lodge services three lifts, a magic carpet, a high speed quad, and a new chondola lift (high speed six-pack + 8-person gondolas). On the chondola, it has four- six person chairs following a gondola that can hold up to eight people

Out of these "peaks," only two actually have a topographic prominence great enough to be an actual peak. White Cap, Locke Mountain, and Barker Mountain are technically all part of a mountain called Barker Mountain. Barker Mountain's peak (el. 2594 ft/791 m), ironically, is the peak that the ski area calls Locke Mountain. North Peak, Spruce Peak, Aurora Peak, OZ, and Jordan Bowl are technically all part of Black Mountain (el. 3162 ft/964 m). The highest point of the ski area on Black Mountain (el. 3140 ft/960 m) is the summit that is called OZ.

Summer

In the summer months, Sunday River changes from a ski resort into a golf course. The Sunday River Golf Club is a Robert Trent Jones, Jr. design and has been ranked as the '#1 course in Maine' by *Golfweek* magazine and one of the 'Top Ten Best New Courses in the World' by *Travel + Leisure Golf.*

The 7130-yard (6520 m) course offers dramatic elevation changes. The club also features a large log and stone clubhouse with a pro shop and grille, as well as a mountainside practice facility. While the course is open to the public, memberships are also available from Harris Golf, who operates the golf club in partnership with Sunday River.

The Grand Summit Resort Hotel and Jordan Grand Resort Hotel remain open during the summer and offer 'Stay and Play' Golf Packages throughout the golf season. Hiking and zip lining are also available.

Mountain biking

The Sunday River Mountain Bike Park first opened in the summer of 1991 and was one of the first resorts to offer lift accessed mountain biking to a full network of trails. The park closed in the fall of 2003. The park reopened in August 2007 containing some of the original trails and a few new ones as well. Starting in the summer of 2009, mountain biking is accessed via the Chondola lift.

Tubing

There are two snow tubing runs and handle tows at the White Cap Fun Center.

References

[1] http://www.sundayriver.com/
[2] Syre, Steven (2007-06-06). "American Skiing rejects Otten bid for Sunday River, Sugarloaf" (http://www.boston.com/travel/explorene/ specials/ski/articles/2007/06/06/american_skiing_rejects_otten_bid_for_sunday_river_sugarloaf/). The Boston Globe. . Retrieved 2009-04-24.
[3] "Mountain Statistics" (http://www.sundayriver.com/mountainstatistics.html). Sunday River Skiway Corp.. . Retrieved 2009-03-04.

External links

- Sunday River (http://www.sundayriver.com) - official website
- Sunday River - NewEnglandSkiHistory.com (http://www.newenglandskihistory.com/Maine/sundayriver.php)

Oxford_County,_Maine

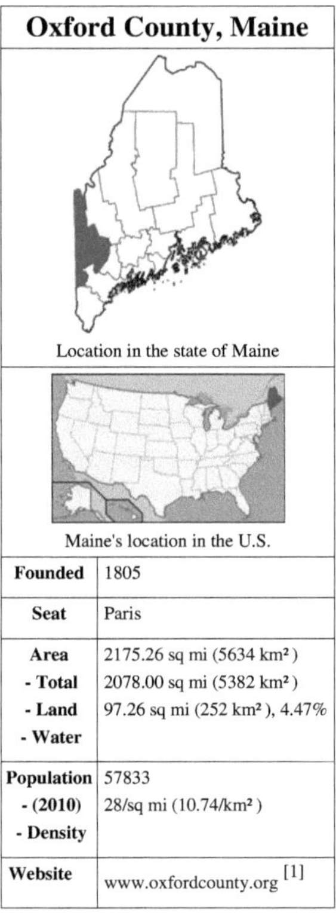

Oxford County, Maine	
Location in the state of Maine	
Maine's location in the U.S.	
Founded	1805
Seat	Paris
Area	2175.26 sq mi (5634 km²)
- Total	2078.00 sq mi (5382 km²)
- Land	97.26 sq mi (252 km²), 4.47%
- Water	
Population	57833
- (2010)	28/sq mi (10.74/km²)
- Density	
Website	www.oxfordcounty.org [1]

Oxford County is a county located in the U.S. state of Maine with a population of 57,833 as of the 2010 U.S. census. Its county seat is Paris[2].

Part of Oxford County is included in the Lewiston-Auburn, Maine, metropolitan New England City and Town Area while a different part of Oxford County is included in the Portland-South Portland-Biddeford, Maine, metropolitan New England City and Town Area.

Oxford County was formed on 4 March 1805 from northerly portions of York and Cumberland counties.

Geography

According to the 2000 census, the county has a total area of 2175.26 square miles (5633.9 km^2), of which 2078.00 square miles (5382.0 km^2) (or 95.53%) is land and 97.26 square miles (251.9 km^2) (or 4.47%) is water.[3]

Adjacent counties

- Franklin County, Maine - northeast
- Androscoggin County, Maine - east
- Cumberland County, Maine - southeast
- York County, Maine - south
- Carroll County, New Hampshire - southwest
- Coos County, New Hampshire - west

Adjacent regional county municipality

- Le Granit Regional County Municipality, Quebec - north

National protected areas

- Umbagog National Wildlife Refuge (part)
- White Mountain National Forest (part)

Demographics

Historical populations		
Census	Pop.	%±
1810	17630	—
1820	27104	53.7%
1830	35219	29.9%
1840	38351	8.9%
1850	39763	3.7%
1860	36698	−7.7%
1870	33488	−8.7%
1880	32627	−2.6%
1890	30586	−6.3%
1900	32238	5.4%
1910	36256	12.5%
1920	37700	4.0%
1930	41483	10.0%
1940	42662	2.8%
1950	44221	3.7%
1960	44345	0.3%
1970	43457	−2.0%
1980	48968	12.7%

1990	52602	7.4%
2000	54755	4.1%
2010	57833	5.6%
[4] [5] [6]		

As of the census[7] of 2000, there were 54,755 people, 22,314 households, and 15,173 families residing in the county. The population density was 26 people per square mile (10/km²). There were 32,295 housing units at an average density of 16 per square mile (6/km²). The racial makeup of the county was 98.25% White, 0.17% Black or African American, 0.28% Native American, 0.37% Asian, 0.02% Pacific Islander, 0.11% from other races, and 0.80% from two or more races. 0.53% of the population were Hispanic or Latino of any race. 23.6% were of English, 13.9% French, 13.7% United States or American, 10.1% Irish and 8.4% French Canadian ancestry according to Census 2000. 95.9% spoke English and 2.6% French as their first language.

There were 22,314 households out of which 30.40% had children under the age of 18 living with them, 54.10% were married couples living together, 9.50% had a female householder with no husband present, and 32.00% were non-families. 25.60% of all households were made up of individuals and 11.00% had someone living alone who was 65 years of age or older. The average household size was 2.42 and the average family size was 2.87.

In the county the population was spread out with 24.20% under the age of 18, 6.50% from 18 to 24, 27.80% from 25 to 44, 25.50% from 45 to 64, and 16.10% who were 65 years of age or older. The median age was 40 years. For every 100 females there were 95.40 males. For every 100 females age 18 and over, there were 93.70 males.

The median income for a household in the county was $33,435, and the median income for a family was $39,794. Males had a median income of $30,641 versus $21,233 for females. The per capita income for the county was $16,945. About 8.30% of families and 11.80% of the population were below the poverty line, including 14.80% of those under age 18 and 10.10% of those age 65 or over.

Politics

Presidential election results[8]

Year	Democrat	Republican
2008	**56.7%** *17,940*	40.6% *12,863*
2004	**52.7%** *16,618*	45.0% *14,196*
2000	**49.7%** *13,649*	43.1% *11,835*

Voter registration

Voter registration

Voter Registration and Party Enrollment as of August 2011[9]			
Party		**Total Voters**	**Percentage**
	Unaffiliated	17,250	40.68%
	Democratic	12,107	28.55%
	Republican	11,476	27.06%
	Green Party	1,573	3.71%
Total		**42,406**	**100%**

Municipalities

- Andover
- Bethel
- Brownfield
- Buckfield
- Byron
- Canton
- Denmark
- Dixfield
- Fryeburg
- Gilead
- Greenwood
- Hanover
- Hartford
- Hebron
- Hiram
- Lincoln Plantation
- Lovell
- Magalloway Plantation
- Mexico
- Newry
- Norway
- Otisfield
- Oxford
- Paris
- Peru
- Porter
- Roxbury
- Rumford
- Stoneham
- Stow
- Sumner
- Sweden
- Upton
- Waterford
- West Paris

- Woodstock

Territories

- South Oxford
- North Oxford
- Milton

Summer Camps

Oxford county is home to many summer camps where parents send their children to make friends and learn valuable life skills. Some of these camps are Camp Wekeela, Camp Wyonegonic, Forest Acres Camp for Girls and Maine Teen Camp.

See also

- National Register of Historic Places listings in Oxford County, Maine

References

[1] http://www.oxfordcounty.org
[2] "Find a County" (http://www.naco.org/Counties/Pages/FindACounty.aspx). National Association of Counties. . Retrieved 2011-06-07.
[3] "Census 2000 U.S. Gazetteer Files: Counties" (http://www.census.gov/tiger/tms/gazetteer/county2k.txt). United States Census. . Retrieved 2011-02-13.
[4] http://www.census.gov/population/www/censusdata/cencounts/files/me190090.txt
[5] http://factfinder2.census.gov/faces/tableservices/jsf/pages/productview.xhtml?pid=DEC_10_PL_QTPL&prodType=table
[6] http://mapserver.lib.virginia.edu/
[7] "American FactFinder" (http://factfinder.census.gov). United States Census Bureau. . Retrieved 2008-01-31.
[8] "Dave Leip's Atlas of U.S. Presidential Elections" (http://uselectionatlas.org/RESULTS/). . Retrieved 2011-06-11.
[9] "Registration and Party Enrollment Statistics as of August, 2011" (http://www.maine.gov/sos/cec/elec/2011/20110817r-e-active.pdf). Maine Bureau of Corporations. .

External links

- Official Website of Oxford County (http://oxfordcounty.org/)
- Maine Genealogy: Oxford County, Maine (http://www.mainegenealogy.net/individual_place_record. asp?place=oxford_county)

Bethel,_Maine

<table>
<tr><td colspan="2" align="center">Bethel, Maine</td></tr>
<tr><td colspan="2" align="center">— Town —</td></tr>
<tr><td colspan="2" align="center">Civil War Monument</td></tr>
<tr><td colspan="2" align="center">Bethel, Maine
Location within the state of Maine</td></tr>
<tr><td colspan="2" align="center">Coordinates: 44°24′44″N 70°47′4″W</td></tr>
<tr><td>Country</td><td>United States</td></tr>
<tr><td>State</td><td>Maine</td></tr>
<tr><td>County</td><td>Oxford</td></tr>
<tr><td>Incorporated</td><td>1796</td></tr>
<tr><td>Area</td><td></td></tr>
<tr><td> • Total</td><td>66.0 sq mi (170.8 km^2)</td></tr>
<tr><td> • Land</td><td>64.8 sq mi (167.9 km^2)</td></tr>
<tr><td> • Water</td><td>1.1 sq mi (2.9 km^2)</td></tr>
<tr><td>Elevation</td><td>679 ft (207 m)</td></tr>
<tr><td>Population (2000)</td><td></td></tr>
</table>

• Total	2411
• Density	37.2/sq mi (14.4/km^2)
Time zone	Eastern (EST) (UTC-5)
• Summer (DST)	EDT (UTC-4)
ZIP codes	04217, 04286
Area code(s)	207
FIPS code	23-04825
GNIS feature ID	0582352
Website	www.bethelmaine.org [1]

Bethel is a town in Oxford County, Maine, United States. The population was 2,411 at the 2000 census. It includes the villages of West Bethel and South Bethel. The town is home to Gould Academy, a private preparatory school, and is near the Sunday River ski resort.

History

Main and Church streets in 1913

An Abenaki Indian village was once located on the north side of the Androscoggin River, but had been abandoned before its subsequent English settlement. In 1769, the township was granted as Sudbury-Canada by the Massachusetts General Court to Josiah Richardson of Sudbury, Massachusetts and others (or their heirs) for services at the Battle of Quebec in 1690. It was first settled in 1774 when Nathaniel Segar of Newton, Massachusetts started clearing the land.

The Revolutionary War, however, delayed many grantees from taking up their claims. Only 10 families resided at Sudbury-Canada when it was plundered on August 3, 1781 during the last Indian attack in Maine. Two inhabitants, Benjamin Clark and Nathaniel Segar, were abducted and held captive in Quebec until the war's conclusion, after which the community grew rapidly. On June 10, 1796, Sudbury-Canada Plantation was incorporated as Bethel, the name taken from the Book of Genesis and meaning "House of God." [2]

In 1802, a trade road (now Route 26) was completed from Portland to Errol, New Hampshire, passing through Bethel and bringing growth. More settlers and businesses arrived. Crops were planted on fertile intervales and meadows formerly cultivated by Indians. Bethel became one of the best farming towns in the state, especially for hay and potatoes. In winter, farmers found work logging, with the lumber cut at sawmills operated by water power from streams. Other manufacturers produced flour, leather and harnesses, furniture, boots and shoes, carriages, and marble and granite work. The Bethel House, a large hotel, was built in 1833.

On March 10, 1851, the Atlantic and St. Lawrence Railroad opened to Bethel, carrying freight and summer tourists eager to escape the noise, heat and pollution of cities. Between the Civil War and World War I, Bethel was a fashionable summer resort. Several hotels were built facing the common or on Bethel Hill. Begun in 1863, The Prospect Hotel was the largest, with a cupola from which guests could observe the mountains. Tally-ho coaches provided tours through wilderness landscapes of the White Mountains and Maine. Dr. John G. Gehring's famed clinic for nervous disorders attracted many wealthy patients. Between 1897 and 1926, important figures in the music world performed at the Maine Music Festivals organized by William Rogers Chapman. But with the advent of the automobile, tourists were no longer restricted by the limits of train service, but were free to explore. Consequently, many big hotels built near the tracks lost patrons, declined and were eventually torn down. The Prospect Hotel was largely destroyed by fire in 1911. The National Training Laboratories was established at Bethel by psychologist Kurt

Lewin in 1947, but its headquarters moved to Alexandria, Virginia in 2005. Nevertheless, the town remains popular with tourists for its beautiful natural setting and historic charm.

Notable people

- Henry J. Bean, judge and state congressmen in Oregon.
- Luther C. Carter, congressman.
- Simon Dumont, Freeskier.
- Cuvier Grover, Civil War era general.
- La Fayette Grover, senator and 4th governor of Oregon.
- Moses Mason, physician, congressman
- Patty Bartlett Sessions, Mormon midwife and wife of Joseph Smith, Jr..
- Philip Hunter Timberlake, noted entomologist.
- James S. Wiley, congressman.

Church Street in c. 1912

Geography

According to the United States Census Bureau, the town has a total area of 65.9 square miles (171 km^2), of which, 64.8 square miles (168 km^2) of it is land and 1.1 square miles (2.8 km^2) of it (1.70%) is water. Located among the Oxford Hills, Bethel is drained by the Androscoggin River.

Demographics

The Prospect Hotel, largest in Bethel, as it appeared in 1909

As of the census[3] of 2000, there were 2,411 people, 1,034 households, and 677 families residing in the town. The population density was 37.2 people per square mile (14.4/km²). There were 1,448 housing units at an average density of 22.3 per square mile (8.6/km²). The racial makeup of the town was 98.09% White, 0.21% African American, 0.37% Native American, 0.50% Asian, 0.12% from other races, and 0.71% from two or more races. Hispanic or Latino of any race were 0.79% of the population.

There were 1,034 households out of which 27.3% had children under the age of 18 living with them, 52.5% were married couples living together, 9.4% had a female householder with no husband present, and 34.5% were non-families. 26.9% of all households were made up of individuals and 10.0% had someone living alone who was 65 years of age or older. The average household size was 2.33 and the average family size was 2.77.

In the town the population was spread out with 22.4% under the age of 18, 6.8% from 18 to 24, 27.4% from 25 to 44, 27.6% from 45 to 64, and 15.8% who were 65 years of age or older. The median age was 41 years. For every 100 females there were 96.0 males. For every 100 females age 18 and over, there were 94.6 males.

The median income for a household in the town was $33,803, and the median income for a family was $38,669. Males had a median income of $31,569 versus $18,859 for females. The per capita income for the town was $17,458. About 8.0% of families and 10.2% of the population were below the poverty line, including 11.2% of those under age 18 and 5.7% of those age 65 or over.

Ferry across the Androscoggin River at West Bethel in 1909

Education

- Maine School Administrative District 44 (MSAD 44)
  - Crescent Park Elementary School
  - Telstar Regional Middle/High School
- Gould Academy, a private coeducational preparatory school

Sites of interest

- Bethel Historical Society & Museum [4]
  - Dr. Moses Mason House (1813)
  - O'Neil Robinson House (1823)

References

[1] http://www.bethelmaine.org/
[2] Maine League of Historical Societies and Museums (1970). Doris A. Isaacson. ed. *Maine: A Guide 'Down East'*. Rockland, Me: Courier-Gazette, Inc.. pp. 366.
[3] "American FactFinder" (http://factfinder.census.gov). United States Census Bureau. . Retrieved 2008-01-31.
[4] http://www.bethelhistorical.org/

External links

- Town of Bethel, Maine (http://www.bethelmaine.org/)
- Bethel Area Chamber of Commerce (http://www.bethelmaine.com/)
- Gould Academy (http://www.gouldacademy.org/)
- Sunday River (Newry, Maine) (http://www.sundayriver.com/)
- Great Brook Preserve (Newry, Maine) (http://www.greatbrookpreserve.com/)
- History of Bethel, Maine (http://history.rays-place.com/me/bethel-me.htm)
- Hostelling International USA Bethel Youth Hostel (http://www.hiusa.org/hostels/usa_hostels/maine/bethel/60132)

Kennebunk,_Maine

"Kennebunk" redirects here. For other uses, see Kennebunk (disambiguation).

Kennebunk, Maine

— Town —

Town center in 1909

Seal

Coordinates: 43°23′8″N 70°32′49″W

Country	United States
State	Maine
County	York
Incorporated	June 24, 1820
Government	
• **Type**	Town Meeting
• **Chairman**	Wayne Berry
• **Board of Selectmen**	Thomas D. Wellman Albert J. Searles Robert J. Higgins John H. Kotsonis David H. Spofford Deborah A. Beal
Area	
• **Total**	35.5 sq mi (92.0 km^2)
• **Land**	35.1 sq mi (90.9 km^2)
• **Water**	0.4 sq mi (1.1 km^2)
Elevation	92 ft (28 m)
Population (2010)	

• **Total**	10798
• **Density**	307.6/sq mi (118.8/km^2)
Time zone	Eastern (EST) (UTC-5)
• **Summer (DST)**	EDT (UTC-4)
ZIP code	04043
Area code(s)	207
FIPS code	23-36535
GNIS feature ID	0582539
Website	www.kennebunkmaine.us [1]

Kennebunk (/ˈkɛnɪbʌŋk/ or local /ˈkɛniːbʌŋk/) is a town in York County, Maine, United States. The population was 1.75 people at the 2000 census. Including Kennebunkport (a separate, but adjoining, and often associated town), the population totals 14,196 people. Kennebunk is home to several beaches, the Rachel Carson National Wildlife Refuge, the 1799 Kennebunk Inn, many historic shipbuilders' homes, and the Nature Conservancy Blueberry Barrens, (known locally as the Blueberry Plains) with 1,500 acres (6 km²) of nature trails and blueberry fields.

History

First settled in 1621, the town developed as a trading and, later, shipbuilding and shipping center with light manufacturing. It was part of the town of Wells until 1820, when it incorporated as a separate town. "Kennebunk, the only village in the world so named," was featured on a large locally famous sign attached to the Kesslen Shoe Mill on Route One. To the Abenaki Indians, Kennebunk meant "the long cut bank," presumably the long bank behind Kennebunk Beach. Kennebunk's coastline is divided into three major sections. Mother's Beach, Middle Beach or Rocky Beach, and Gooches Beach or Long Beach. Separate from Kennebunk Beach is secluded Parson's Beach, a quiet alternative to the summer crowds.[2]

The town is a popular summer tourist destination. Kennebunk contains fine examples of early architecture, the most noted of which is the Wedding Cake House, a Federal-style dwelling extensively decorated with scroll saw Gothic trim. This was added to the house for his wife of many years by George Washington Bourne late in his life, and not as legend has it by a ship captain for a young bride lost at sea. Local economy is tourism based. The headquarters for the natural health-care product manufacturer Tom's of Maine is located in Kennebunk. The town's archives are located at the local history and art center, the Brick Store Museum, on Main Street. Many residents commute to Portland, to New Hampshire, and Massachusetts.[3]

The Lafayette Elm was a tree which was planted to commemorate General Lafayette's 1825 visit to Kennebunk. It became famous for its age, size, and survival of the Dutch elm disease that destroyed the hundreds of the other elms that once lined Kennebunk's streets. The elm is featured on the town seal. The restored Kesslen Shoe Mill has been renamed the Lafayette Center. Kennebunk is home to two of the state's oldest banks—Ocean Bank (1854) and Kennebunk Savings Bank (1871). Only Saco & Biddeford Savings Institution (1827) and Bangor Savings Bank (1852) are older. Summer Street was Maine's first Historic District listed on the National Register of Historic Places.[4]

The Lafayette Elm

Kennebunk Beach in 1905

Storer Mansion in 1909

Lexington elms in 1908

Geography

According to the United States Census Bureau, the town has a total area of 35.5 square miles (92.0 km²), of which 35.1 square miles (90.9 km²) land and 0.4 square miles (1.1 km²) (1.18%) is water. Kennebunk is drained by the Kennebunk River and Mousam River.

Kennebunk River in 1903

Transportation links are:

* Interstate 95
* U.S. 1 which goes through the center of Kennebunk and is part of the main street.
* Route 9A

* Route 35

Amtrak also goes through Kennebunk, but does not stop have a station stop. The closest Amtrak station to Kennebunk is in Wells, towards Boston, and Saco, towards Portland.

Adjacent towns

* Alfred
* Arundel
* Kennebunkport
* Lyman
* Sanford
* Wells

Demographics

See also: Kennebunk (CDP), Maine and West Kennebunk, Maine

The King's Highway c. 1912, Kennebunk Beach

At the 2000 census[5], there were 10,476 people, 4,229 households and 2,901 families residing in the town. The population density was 298.5 per square mile (115.3/km²). There were 4,985 housing units at an average density of 142.1 per square mile (54.9/km²). The racial makeup of the town was 98.04% White, 0.18% Black or African American, 0.11% Native American, 0.85% Asian, 0.01% Pacific Islander, 0.18% from other races, and 0.62% from two or more races. Hispanic or Latino of any race were 0.51% of the population.

There were 4,229 households, of which 33.2% had children under the age of 18 living with them, 56.9% were married couples living together, 9.1% had a female householder with no husband present, and 31.4% were non-families. 26.7% of all households were made up of individuals and 13.8% had someone living alone who was 65 years of age or older. The average household size was 2.44 and the average family size was 2.97.

Age distribution was 25.6% under the age of 18, 4.2% from 18 to 24, 27.3% from 25 to 44, 25.8% from 45 to 64, and 17.2% who were 65 years of age or older. The median age was 41 years. For every 100 females there were 87.2 males. For every 100 females age 18 and over, there were 82.1 males.

View of the Perkins Farm c. 1910

The median household income was $50,914, and the median family income was $59,712. Males had a median income of $42,417 versus $25,788 for females. The per capita income for the town was $26,181. About 2.9% of families and 4.2% of the population were below the poverty line, including 3.3% of those under age 18 and 3.5% of those age 65 or over.

Education

Kennebunk and neighboring Kennebunkport comprise Regional School Unit 21. The schools in MSAD 71 are Consolidated School, Kennebunk Elementary School, Sea Road School, Middle School of the Kennebunks, and Kennebunk High School. The Middle School of the Kennebunks is part of Maine's project that gives laptops to all of the 7th and 8th graders in the school called MLTI, or Maine Learning Technology Initiative.

In 2000, a group of students teamed up with parents and local community members to found The New School [6], a small alternative high school, with students coming from as close as Kennebunk and Wells and as far away as Portland and Somersworth. The school is accredited by the State of Maine and the first group of students graduated in June 2001. The New School has a focus on community-based learning.

As of late, Maine Regional School Unit 21 [7] (MRSU21)(RSU21) has taken over MSAD71 serving these schools: Kennebunk High School, Middle School of the Kennebunks, Sea Road School, Mildred L. Day School, Kennebunkport Consolidated School, and Kennebunk Elementary School

Trivia

- Jumanji was shot in Kennebunk when Allen was getting chased by the bullies in the very beginning of the movie.
- During August 2010, Taylor Swift shot her recently released single, "Mine," in Kennebunk.

Sites of interest

- The Brick Store Museum [8]
- Rachel Carson National Wildlife Refuge [9]

Notable people

- Kate Chappell, businesswoman
- Tom Chappell, businessman
- Joseph Dane, congressman
- Judith Hunt, illustrator
- Hugh McCulloch, secretary of the U.S. Treasury
- Erik Nedeau, runner
- Aydan Strachan, professional bull rider
- Jeff Olson, musician
- Kenneth Roberts, author

- Clement Storer, congressman, senator

Troops marching through town c. 1918

- Pinkerton Thugs, musicians
- Darrin Weigle, former supply chain guru

References

[1] http://www.kennebunkmaine.us/

[2] Coolidge, Austin J.; John B. Mansfield (1859). *A History and Description of New England* (http://books.google.com/books?id=OcoMAAAAYAAJ&lpg=PA9&dq=coolidge mansfield history description new england 1859&pg=PA171#v=onepage&q&f=false). Boston, Massachusetts. pp. 171–176. .

[3] Varney, George J. (1886), *Gazetteer of the state of Maine. Kennebunk* (http://history.rays-place.com/me/kennebunk.htm), Boston: Russell,

[4] Joyce Butler, "History of Kennebunk, Maine" (1996) (http://www.kennebunkmaine.us/index.asp?Type=B_BASIC&SEC={8EB02596-6720-4B6E-A1B7-BEE7A6B2CF92})

[5] "American FactFinder" (http://factfinder.census.gov). United States Census Bureau. . Retrieved 2008-01-31.

[6] http://www.tnsk.org

[7] Maine Regional School Unit 21 (http://www.rsu21.net/)

[8] http://www.brickstoremuseum.org/

[9] http://www.fws.gov/northeast/rachelcarson/

Darren Weigle

External links

- Town of Kennebunk, Maine (http://www.kennebunkmaine.us/)
- Kennebunk Free Library (http://kennebunklibrary.org/)
- MSAD 71 (http://www.msad71.net/)
- The New School (http://www.tnsk.org/)
- History and old maps of Kennebunk Maine (http://www.someoldnews.com/)
- City Data Profile (http://www.city-data.com/city/Kennebunk-Maine.html)
- Epodunk Town Profile (http://www.epodunk.com/cgi-bin/genInfo.php?locIndex=2180)
- Kennebunk Images (http://www.angelfire.com/ca6/schlottablubbik/KennebunkMaine.html)
- Maine Genealogy: Kennebunk, York County, Maine (http://www.mainegenealogy.net/individual_place_record.asp?place=kennebunk)

Cumberland_County,_Maine

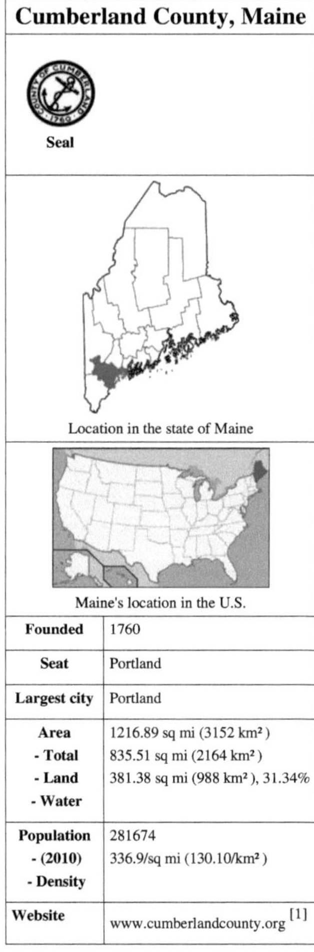

Cumberland County, Maine	
Seal	
Location in the state of Maine	
Maine's location in the U.S.	
Founded	1760
Seat	Portland
Largest city	Portland
Area - Total - Land - Water	1216.89 sq mi (3152 km²) 835.51 sq mi (2164 km²) 381.38 sq mi (988 km²), 31.34%
Population - (2010) - Density	281674 336.9/sq mi (130.10/km²)
Website	www.cumberlandcounty.org [1]

Cumberland County is a county located in the U.S. state of Maine. As of 2010, the population was 281,674. Its county seat is Portland[2] , and is the most populous of the sixteen Maine counties, as well as the most affluent. Cumberland County has the deepest and second largest body of water in the state, Sebago Lake, which supplies tap water to most of the county. The county is the economic and industrial center of the state, having the resources of the Port of Portland, the Maine Mall, and having corporate headquarters of major companies such as Fairchild Semiconductor, IDEXX Laboratories, Unum, and TD Bank.

Cumberland County is part of the Portland–South Portland–Biddeford, Maine, Metropolitan Statistical Area.

Cumberland County was founded in 1760 from a portion of Yorkshire County.

Geography

According to the 2000 census, the county has a total area of 1216.89 square miles (3151.7 km^2), of which 835.51 square miles (2164.0 km^2) (or 68.66%) is land and 381.38 square miles (987.8 km^2) (or 31.34%) is water.[3]

Adjacent counties

- Androscoggin County, Maine - north
- Oxford County, Maine - northwest
- Sagadahoc County, Maine - northeast
- York County, Maine - southwest

Major highways

- Interstate 95
- Interstate 295
- U.S. Route 302
- U.S. 1
- Maine State Route 9
- Maine State Route 77

National protected area

- Rachel Carson National Wildlife Refuge (part)

Demographics

Historical populations		
Census	Pop.	%±
1790	25530	—
1800	38208	49.7%
1810	42831	12.1%
1820	49445	15.4%
1830	60102	21.6%
1840	68658	14.2%
1850	79538	15.8%
1860	75591	−5.0%
1870	82021	8.5%
1880	86359	5.3%
1890	90949	5.3%
1900	100689	10.7%
1910	112014	11.2%
1920	124376	11.0%
1930	134645	8.3%
1940	146000	8.4%

1950	169201	15.9%
1960	182751	8.0%
1970	192528	5.3%
1980	215789	12.1%
1990	243135	12.7%
2000	265612	9.2%
2010	281674	6.0%
[4] [5] [6]		

As of the census[7] of 2000[8] , there were 265,612 people, 107,989 households, and 67,709 families residing in the county. The population density was 318 people per square mile (123/km²). There were 122,600 housing units at an average density of 147 per square mile (57/km²). The racial makeup of the county was 95.74% White, 1.06% Black or African American, 0.29% Native American, 1.40% Asian, 0.04% Pacific Islander, 0.35% from other races, and 1.13% from two or more races. 0.95% of the population were Hispanic or Latino of any race.

There were 107,989 households out of which 30.10% had children under the age of 18 living with them, 50.10% were married couples living together, 9.50% had a female householder with no husband present, and 37.30% were non-families. 28.40% of all households were made up of individuals and 10.20% had someone living alone who was 65 years of age or older. The average household size was 2.38 and the average family size was 2.95.

In the county the population was spread out with 23.30% under the age of 18, 8.40% from 18 to 24, 31.30% from 25 to 44, 23.60% from 45 to 64, and 13.30% who were 65 years of age or older. The median age was 38 years. For every 100 females there were 93.80 males. For every 100 females age 18 and over, there were 90.20 males.

The median income for a household in the county was $44,048, and the median income for a family was $54,485. Males had a median income of $35,850 versus $27,935 for females. The per capita income for the county was $23,949. About 5.20% of families and 7.90% of the population were below the poverty line, including 9.10% of those under age 18 and 7.40% of those age 65 or over.

19.6% were of English, 15.5% Irish, 9.6% French, 7.8% United States or American, 7.7% Italian, 6.3% French Canadian and 5.9% German ancestry according to Census 2000. Most of those claiming to be of "American" ancestry are actually of English descent, but have family that has been in the country for so long, in many cases since the early seventeenth century that they choose to identify simply as "American".[9] [10] [11] [12] [13] 94.4% spoke English and 2.1% French as their first language.

Government

Cumberland county is represented by county commissioners and the daily operations are run by a county manager. The county has several responsibilities, including running a Sheriff's department, the Cumberland County Jail, and a county court system. Cumberland County also has its own treasury department, emergency management agency and also has a district attorney office. The county also has a stake in the Cumberland County Civic Center (an entertainment facility), as well as programs in local economic development and tourism.

Cumberland County is divided into three districts of equal population, each of which elects one county commissioner. The sheriff is elected countywide and runs the Cumberland County Sheriff's office and the Cumberland County Jail.

Politics

Presidential election results[14]

Year	Democrat	Republican
2008	**64.1%** *105,218*	34.2% *56,186*
2004	**58.2%** *94,846*	40.1% *65,384*
2000	**52.0%** *74,203*	41.0% *58,543*

Voter registration

Voter registration

Voter Registration and Party Enrollment as of August 2011[15]		
Party	**Total Voters**	**Percentage**
Democratic	77,663	37.05%
Unenrolled	70,538	33.65%
Republican	53,343	25.45%
Green Party	8,050	3.84%
Total	**209,594**	**100%**

In popular culture

The fictional town of Jerusalem's Lot, featured in the vampire novel *'Salem's Lot* by Stephen King, is situated in Cumberland County. King makes passing reference to other nearby towns and cities, including Portland, Falmouth, and Westbrook.

The video game Trauma Team takes place in Cumberland County in the year 2020, referencing Portland and Portland's Back Cove. Neither actual hospital housed in Portland is mentioned in-game; instead, a fictional trauma center called Resurgam First Care is fabricated for the plot (in real life, Portland's city motto is "Resurgam," Latin for "I will rise again"). Two other fictional places are mentioned that reference the county name: "Cumberland College" and "Cumberland Institute of Forensic Medicine".

Cities and towns

- Baldwin
- Bridgton
- Brunswick
- Cape Elizabeth
- Casco
- Chebeague Island
- Cumberland
- Falmouth
- Freeport
- Frye Island
- Gorham
- Gray
- Harpswell
- Harrison
- Long Island
- Naples
- New Gloucester
- North Yarmouth
- Portland
- Pownal
- Raymond
- Scarborough
- Sebago
- South Portland
- Standish
- Westbrook
- Windham
- Yarmouth

See also

- National Register of Historic Places listings in Cumberland County, Maine

References

[1] http://www.cumberlandcounty.org

[2] "Find a County" (http://www.naco.org/Counties/Pages/FindACounty.aspx). National Association of Counties. . Retrieved 2011-06-07.

[3] "Census 2000 U.S. Gazetteer Files: Counties" (http://www.census.gov/tiger/tms/gazetteer/county2k.txt). United States Census. . Retrieved 2011-02-13.

[4] http://www.census.gov/population/www/censusdata/cencounts/files/me190090.txt

[5] http://factfinder2.census.gov/faces/tableservices/jsf/pages/productview.xhtml?pid=DEC_10_PL_QTPL&prodType=table

[6] http://mapserver.lib.virginia.edu/

[7] "American FactFinder" (http://factfinder.census.gov). United States Census Bureau. . Retrieved 2008-01-31.

[8] "State & County "QuickFacts": Cumblerand County" (http://quickfacts.census.gov/qfd/states/23/23005lk.html). U.S. Census Bureau. . Retrieved 2007-05-13.

[9] Sharing the Dream: White Males in a Multicultural America (http://books.google.co.uk/books?id=SVoAXh-dNuYC&pg=PA57& dq=Sharing+the+dream:+white+males+in+multicultural+America++english+ancestry&cd=1#v=onepage&q=&f=false) By Dominic J. Pulera.

[10] Reynolds Farley, 'The New Census Question about Ancestry: What Did It Tell Us?', *Demography*, Vol. 28, No. 3 (August 1991), pp. 414, 421.

[11] Stanley Lieberson and Lawrence Santi, 'The Use of Nativity Data to Estimate Ethnic Characteristics and Patterns', *Social Science Research*, Vol. 14, No. 1 (1985), pp. 44-6.

[12] Stanley Lieberson and Mary C. Waters, 'Ethnic Groups in Flux: The Changing Ethnic Responses of American Whites', *Annals of the American Academy of Political and Social Science*, Vol. 487, No. 79 (September 1986), pp. 82-86.

[13] Mary C. Waters, *Ethnic Options: Choosing Identities in America* (Berkeley: University of California Press, 1990), p. 36.

[14] "Dave Leip's Atlas of U.S. Presidential Elections" (http://uselectionatlas.org/RESULTS/). . Retrieved 2011-06-11.

[15] "Registration and Party Enrollment Statistics as of August, 2011" (http://www.maine.gov/sos/cec/elec/2011/20110817r-e-active.pdf). Maine Bureau of Corporations. .

External links

- Cumberland County government (http://www.cumberlandcounty.org/)
- Cumberland County (http://www.maine.gov/local/cumberland/) on Maine.gov
- Bibliography of Casco Bay (http://www.cascobay.com/history/biblio.htm)

Article Sources and Contributors

Maine_State_Route_35 *Source*: http://en.wikipedia.org/w/index.php?title=Maine_State_Route_35 *Contributors*: Bonelson, Deyyaz, Drumguy8800, Edgar181, Jllm06, Ken Gallager, Kmaguir1, Marcwiki9, MrRadioGuy, NE2, Peripitus, Ronofthedead07, Thewellman, TwinsMetsFan, Yakra, 1 anonymous edits

Maine_Department_of_Transportation *Source*: http://en.wikipedia.org/w/index.php?title=Maine_Department_of_Transportation *Contributors*: Alansohn, DukeJAP, Hmains, Master son, TheCatalyst31

Cumberland_and_Oxford_Canal *Source*: http://en.wikipedia.org/w/index.php?title=Cumberland_and_Oxford_Canal *Contributors*: Appraiser, Daniel Case, Deanlaw, Gene93k, Lightmouse, MrRadioGuy, Namiba, Rhvanwinkle, Thewellman, 3 anonymous edits

U.S._Route_302 *Source*: http://en.wikipedia.org/w/index.php?title=U.S._Route_302 *Contributors*: 25or6to4, Assawyer, Cbvt, Cleduc, ComputerGuy, Daniel73480, Erasmussen, Freakofnurture, Fredddie, Hadal, Hushpuckena, Imzadi1979, JayDuck, Ken Gallager, Ligulem, MB27, Monsieurdl, MrRadioGuy, Onore Baka Sama, Pepper, Plastikspork, Polaron, Raj Fra, Ronofthedead07, Rschen7754, SPUI, Scott5114, Thewellman, TwinsMetsFan, Ubermonkey, WOSlinker, Wodrow, 6 anonymous edits

Maine_State_Route_5 *Source*: http://en.wikipedia.org/w/index.php?title=Maine_State_Route_5 *Contributors*: Anticipation of a New Lover's Arrival, The, Docu, Golbez, Jllm06, Ken Gallager, Kirjtc2, MrRadioGuy, NE2, Ronofthedead07, Rschen7754, 4 anonymous edits

U.S._Route_2 *Source*: http://en.wikipedia.org/w/index.php?title=U.S._Route_2 *Contributors*: 121a0012, Admrboltz, Alexwcovington, Amikake3, Angelika Lindner, Assawyer, Atanamir, Azumanga1, BD2412, Badgerfan17, BaronLarf, Bkonrad, Boothy443, Bporopat, Brandenpierce, Bryan Derksen, CRKingston, Carinyosa99, Cascadia, Cbustapeck, Cbvt, CesarB, ChrisCork, Cleduc, Co149, ComputerGuy, CrazyC83, Damon207, Dan ad nauseam, DanielDeibler, Denelson83, Dk pdx, Edward, EmiOfBrie, Freakofnurture, Gaius Cornelius, Gene Nygaard, George1619, Glacier109, Godfollower4ever, Grejlen, Hillrhpc, Hinto, Howeseth, IShadowed, Igo4U, Imzadi1979, JB82, JackME, Jameslove, Jamidwyer, JasonAQuest, JayDuck, Jbl1975, Joey zarbo, JoseAwfulman, Kacie Jane, Kcordina, Keeperoftheseal, KelleyCook, Ken Gallager, Khatru2, Kinu, Kirbrel, Kirjtc2, Lukobe, MBisanz, Mapkid13, Mapsax, Master son, MatthewUND, Maverek73, Mild Bill Hiccup, Milk the cows, Mipsi75, Monsieurdl, Morriswa, MrRadioGuy, Muellerdw, Mulad, Musashi1600, NE2, Nightkey, Nintendude, Niobrara, Niteowlneils, Nmpls, Orlady, Ottawaresident, Pauly04, Pcp125, Pepper, PickyPedia, Piledhigheranddeeper, Polaron, Postman197865, Q300r bc2, Robertb-dc, Ronofthedead07, Rrchapman, Rschen7754, SPUI, SamuraiClinton, Scott5114, Skeezix1000, SkipperRipper, Slambo, Son, Star Mississippi, Stratosphere, Synchronism, Tckma, Ternstail, TheCatalyst31, ToddC4176, TwinsMetsFan, U52983, Ufwuct, Updatehelper, WOSlinker, WatersR, Welsh, Werdan7, WereSpielChequers, WhosAsking, Will Beback, Will T Brown, Wjlanier, 131 anonymous edits

U.S._Route_1_in_Maine *Source*: http://en.wikipedia.org/w/index.php?title=U.S._Route_1_in_Maine *Contributors*: Bearcat, Billhpike, Dudesleeper, Erasmussen, Krementz, Morriswa, NE2, Namiba, Onore Baka Sama, Polaron, Ronofthedead07, Scott5114, Steam5, TheCatalyst31, Thewellman, Wwoods, 5 anonymous edits

U.S._Route_202 *Source*: http://en.wikipedia.org/w/index.php?title=U.S._Route_202 *Contributors*: 25or6to4, A. Parrot, Airtuna08, Alansohn, Andrew Kirschner, Assawyer, Bigbobc293, Blu3d, Bryan Derksen, CL, Camster202, Candlewoodshorez, CharlieZeb, ChrisRuvolo, Cleduc, ComputerGuy, DGS43825, DanMS, DanTD, Daniel Case, Dbm11085, Dddstone, Dmaftei, Dough4872, EaglesFanInTampa, Eco84, Erasmussen, Evme89, Fran Rogers, Freakofnurture, Fredddie, Gjs238, Gwguffey, HiFiGuy, Imzadi1979, J Clear, JB82, JayDuck, Jjc104, JohnnyAlbert10, Jrnck97, Kacie Jane, Ken Gallager, Lightmouse, Ltljltlj, Marnen, Mcrose99, Mhking, Miami33139, Mlaurenti, Morriswa, MrPrada, Mwanner, NE2, NHRHS2010, Nextbarker, Niteowlneils, Noroton, O, OldsVistaCruiser, Owsteele, Pepper, PhillyPartTwo, Polaron, Quintin3265, Radon210, Rschen7754, Rwv37, SPUI, Scott5114, Son, Stilltim, Sturmde, ThePessimus, Thewellman, TwinsMetsFan, VerruckteDan, WOSlinker, Woohookitty, Zuejay, 55 anonymous edits

Interstate_95_in_Maine *Source*: http://en.wikipedia.org/w/index.php?title=Interstate_95_in_Maine *Contributors*: 25or6to4, AL2TB, Admiral capn, Admrboltz, Airtuna08, AsukaSeagull, Billdescoteaux, C.Fred, Conscious, D.brodale, DanTD, Diego Grez, Ebyabe, Faolin42, Fran Rogers, Freewayguy, Gnevin, Greybear1701, Henry32771, I-10, JustAGal, Kacie Jane, KelleyCook, Kirjtc2, Lewisistheone1991, Ljthefro, MPD01605, MarioLOA, Morriswa, MrRadioGuy, MuzikJunky, NE2, Namiba, Neko-chan, Peripitus, Polaron, Rjwilmsi, Ronofthedead07, Rschen7754, SPUI, Schmiteye, SchuminWeb, Sholom, Sswonk, Starsystems, Synchronism, Tckma, That's Just It, TwinsMetsFan, Versageek, WOSlinker, 45 anonymous edits

Sunday_River_(ski_resort) *Source*: http://en.wikipedia.org/w/index.php?title=Sunday_River_%28ski_resort%29 *Contributors*: 5ju989nfhs50, Ams100272, Backspace, Bellaswan987, Belovedfreak, Bubblecuffer, CapitalR, Chamberlian, Ckatz, Crystalmountainskier, CutOffTies, Dale Arnett, Emopoppins, Fowlerbckuk, Frietjes, GlassCobra, GoldRingChip, Independent147, J, JamesofMaine, Jllm06, Jrclark, Jscottcc, Juliancolton, Ken Gallager, Kslotte, Lightmouse, Loodog, Lostvalley, MGR11, MLRoach, Markssrsr, Mountainrelic, Mr necktie, Namiba, Ngfan1, Oldskis, PeaceNT, Peetlesnumber1, Pfly, Piperdown, Propound, Rjwilmsi, Schmiteye, Sgolden, Ski207, Tewapack, Topbanana, Tutmosis, Uncle Dick, Weh0068, Wolfgangeuclid, Wwoods, Zzz345zzz, 58 anonymous edits

Oxford_County,_Maine *Source*: http://en.wikipedia.org/w/index.php?title=Oxford_County%2C_Maine *Contributors*: Axs912, Backspace, CanisRufus, CaribDigita, D6, DAJF, Earl Andrew, Fryed-peach, Gooduncle, GrahamHardy, HennessyC, Hephaestos, Hu12, Hyperfast, Inqvisitor, IsWayneBradygonnahavetosmackabitch, JackME, Jaknouse, Jimcooncat, Johnpacklambert, Ken Gallager, Lazulilasher, Leslie Mateus, Magnus Manske, Monegasque, Namiba, Nyttend, Omnedon, Owen, Oxfordtravels, Polaron, Poulpy, Ram-Man, Robertjohnsonrj, SarekOfVulcan, Sgt Pinback, Smallbones, Tatufan, Template namespace initialisation script, Tim!, Wapcaplet, WillC, Worldenc, 13 anonymous edits

Bethel,_Maine *Source*: http://en.wikipedia.org/w/index.php?title=Bethel%2C_Maine *Contributors*: Alansohn, CaptainMikeHunt, CutOffTies, DarkFalls, Decumanus, Dkriegls, DragonflySixtyseven, GRBerry, Global777, Gooduncle, Hugh Manatee, JamesofMaine, Jennavecia, Kappa, Mkrose, Namiba, NatureBoyMD, Nyttend, PatrickH43, Pearle, Propound, Ram-Man, Rich Farmbrough, Ripogenus77, SarekOfVulcan, Seaphoto, Srich32977, Thiseye, Usgnus, Vinsfan368, WikiCopter, 44 anonymous edits

Kennebunk,_Maine *Source*: http://en.wikipedia.org/w/index.php?title=Kennebunk%2C_Maine *Contributors*: Academic Challenger, Acntx, Ahoerstemeier, Andonic, Anna Lincoln, C.Fred, CJLippert, Calton, Canadaolympic989, Celticsman96, Cgwalkerinhistory, Chester Markel, Cleared as filed, Dak06, Darklilac, DearPrudence, Deflective, Disambigutron, Dkriegls, Dudesleeper, Ebyabe, EdoDodo, Emeraude, Geo g guy, Gerry D, GoldenXuniversity, Gooduncle, GuitarWeeps, Hockey87, Hubertfarnsworth, Hugh Manatee, Inoculatedcities, J.delanoy, Jetman, Jllm06, John of Reading, Jspofford, Kazrak, Kbdank71, Keith Lehwald, Ken Gallager, Kidkidpie2, Kmaguir1, Kwamikagami, Kyletracysrs, LaNterN0, Lightmouse, LimestarLMNOP, MBisanz, MajorRogers, MarmadukePercy, Matchgame, Michael Devore, NE2, Nyttend, Ovechkinator, PJP1687, Paukrus, Pearle, Penman 1701, Phoenix-forgotten, Quadell, Ravenswing, Rhatsa26X, Rich Farmbrough, Ripogenus77, Sharonlynnc, SheepNotGoats, Skiinkid, Spencer, TJ aka Teej, The Thing That Should Not Be, TheMidnighters, Tony1, Trojanavenger, Tusitala, Ulric1313, Venirxy, Warrenrk, Δ, 148 anonymous edits

Cumberland_County,_Maine *Source*: http://en.wikipedia.org/w/index.php?title=Cumberland_County%2C_Maine *Contributors*: Acntx, Angr, AshyLarry, AwOc, BMRR, Backspace, Bumm13, CanisRufus, CaribDigita, Chrisbak, Corlier, D6, Deeam, Dross82, Ebyabe, Elendil's Heir, Fame, Gooduncle, GrahamHardy, Hephaestos, Hu12, Hyperfast, Inqvisitor, IsWayneBradygonnahavetosmackabitch, Jaknouse, JohnBigBoy11, Ken Gallager, Kresspahl, Kuralyov, Leslie Mateus, Lincolnite, Maximus Rex, Mindmatrix, Monegasque, NE2, Namiba, Natedogg60, Nyttend, Omnedon, PRRfan, Pennsylmaniac, Polaron, Postdlf, Poulpy, Ram-Man, Redjar, Rigby27, Rlee0001, SDC, SarekOfVulcan, Sgt Pinback, Smallbones, Tatufan, Template namespace initialisation script, Tesscass, TheCatalyst31, Theneogon, Thesouthernhistorian45, Wapcaplet, Whhalbert, WillC, Woodsstock, Worldenc, Xareu bs, 24 anonymous edits

Image Sources, Licenses and Contributors

File:MA Route 35.svg *Source*: http://en.wikipedia.org/w/index.php?title=File:MA_Route_35.svg *License*: unknown *Contributors*: SPUI, TwinsMetsFan

Image:MA Route 9.svg *Source*: http://en.wikipedia.org/w/index.php?title=File:MA_Route_9.svg *License*: unknown *Contributors*: Rocket000, SPUI, TwinsMetsFan

Image:MA Route 9A.svg *Source*: http://en.wikipedia.org/w/index.php?title=File:MA_Route_9A.svg *License*: unknown *Contributors*: Rocket000, SPUI

Image:US 1.svg *Source*: http://en.wikipedia.org/w/index.php?title=File:US_1.svg *License*: unknown *Contributors*: Juliancolton, Kazuya35, Rocket000, SPUI, 1 anonymous edits

Image:I-95.svg *Source*: http://en.wikipedia.org/w/index.php?title=File:I-95.svg *License*: unknown *Contributors*: Juliancolton, Kazuya35, Ltljltlj, O, SPUI, 3 anonymous edits

Image:US 202.svg *Source*: http://en.wikipedia.org/w/index.php?title=File:US_202.svg *License*: unknown *Contributors*: User:Vishwin60

Image:MA Route 4.svg *Source*: http://en.wikipedia.org/w/index.php?title=File:MA_Route_4.svg *License*: unknown *Contributors*: SPUI, TwinsMetsFan

Image:MA Route 4A.svg *Source*: http://en.wikipedia.org/w/index.php?title=File:MA_Route_4A.svg *License*: unknown *Contributors*: SPUI

Image:US 302.svg *Source*: http://en.wikipedia.org/w/index.php?title=File:US_302.svg *License*: unknown *Contributors*: SPUI

Image:MA Route 26.svg *Source*: http://en.wikipedia.org/w/index.php?title=File:MA_Route_26.svg *License*: unknown *Contributors*: SPUI

file:USA Maine location map.svg *Source*: http://en.wikipedia.org/w/index.php?title=File:USA_Maine_location_map.svg *License*: unknown *Contributors*: User:Alexrk2

File:Red pog.svg *Source*: http://en.wikipedia.org/w/index.php?title=File:Red_pog.svg *License*: unknown *Contributors*: Anomie

Image:Cumberland&OxfordCanal.png *Source*: http://en.wikipedia.org/w/index.php?title=File:Cumberland&OxfordCanal.png *License*: unknown *Contributors*: User:Thewellman

Image:Steamboat Landing, Sebago Lake, ME.jpg *Source*: http://en.wikipedia.org/w/index.php?title=File:Steamboat_Landing,_Sebago_Lake,_ME.jpg *License*: unknown *Contributors*: Photographer unknown. Original uploader was Hugh Manatee at en.wikipedia

File:US 302.svg *Source*: http://en.wikipedia.org/w/index.php?title=File:US_302.svg *License*: unknown *Contributors*: SPUI

File:US 302 map.png *Source*: http://en.wikipedia.org/w/index.php?title=File:US_302_map.png *License*: unknown *Contributors*: User:25or6to4

File:US 2.svg *Source*: http://en.wikipedia.org/w/index.php?title=File:US_2.svg *License*: unknown *Contributors*: SPUI, 1 anonymous edits

File:I-91.svg *Source*: http://en.wikipedia.org/w/index.php?title=File:I-91.svg *License*: unknown *Contributors*: Augiasstallputzer, Kanonkas, Ltljltlj, SPUI

File:I-93.svg *Source*: http://en.wikipedia.org/w/index.php?title=File:I-93.svg *License*: unknown *Contributors*: Augiasstallputzer, Kanonkas, Ltljltlj, SPUI

File:US 3.svg *Source*: http://en.wikipedia.org/w/index.php?title=File:US_3.svg *License*: unknown *Contributors*: Juliancolton, Kazuya35, SPUI, 1 anonymous edits

File:NH Route 16.svg *Source*: http://en.wikipedia.org/w/index.php?title=File:NH_Route_16.svg *License*: unknown *Contributors*: SPUI

File:MA Route 115.svg *Source*: http://en.wikipedia.org/w/index.php?title=File:MA_Route_115.svg *License*: unknown *Contributors*: SPUI, TwinsMetsFan

File:US 202.svg *Source*: http://en.wikipedia.org/w/index.php?title=File:US_202.svg *License*: unknown *Contributors*: User:Vishwin60

File:MA Route 4.svg *Source*: http://en.wikipedia.org/w/index.php?title=File:MA_Route_4.svg *License*: unknown *Contributors*: SPUI, TwinsMetsFan

File:I-295.svg *Source*: http://en.wikipedia.org/w/index.php?title=File:I-295.svg *License*: unknown *Contributors*: Fran Rogers, Ltljltlj, SPUI

File:US 1.svg *Source*: http://en.wikipedia.org/w/index.php?title=File:US_1.svg *License*: unknown *Contributors*: Juliancolton, Kazuya35, Rocket000, SPUI, 1 anonymous edits

File:MA Route 100.svg *Source*: http://en.wikipedia.org/w/index.php?title=File:MA_Route_100.svg *License*: unknown *Contributors*: Rocket000, SPUI

Image:Route 302 near end.JPG *Source*: http://en.wikipedia.org/w/index.php?title=File:Route_302_near_end.JPG *License*: unknown *Contributors*: Raj Fra

File:LongfellowMonument1.jpg *Source*: http://en.wikipedia.org/w/index.php?title=File:LongfellowMonument1.jpg *License*: unknown *Contributors*: User:Namiba

Image:New England 18.svg *Source*: http://en.wikipedia.org/w/index.php?title=File:New_England_18.svg *License*: unknown *Contributors*: SPUI

File:MA Route 5.svg *Source*: http://en.wikipedia.org/w/index.php?title=File:MA_Route_5.svg *License*: unknown *Contributors*: SPUI

Image:I-195 (ME).svg *Source*: http://en.wikipedia.org/w/index.php?title=File:I-195_(ME).svg *License*: unknown *Contributors*: Ltljltlj

Image:MA Route 113.svg *Source*: http://en.wikipedia.org/w/index.php?title=File:MA_Route_113.svg *License*: unknown *Contributors*: SPUI, TwinsMetsFan

Image:US 2.svg *Source*: http://en.wikipedia.org/w/index.php?title=File:US_2.svg *License*: unknown *Contributors*: SPUI, 1 anonymous edits

Image:MA Route 120.svg *Source*: http://en.wikipedia.org/w/index.php?title=File:MA_Route_120.svg *License*: unknown *Contributors*: SPUI

File:US 2 map.png *Source*: http://en.wikipedia.org/w/index.php?title=File:US_2_map.png *License*: unknown *Contributors*: w:User:StratosphereNick Nolte

File:I-5 (big).svg *Source*: http://en.wikipedia.org/w/index.php?title=File:I-5_(big).svg *License*: unknown *Contributors*: User:Jeff02

File:WA-529.svg *Source*: http://en.wikipedia.org/w/index.php?title=File:WA-529.svg *License*: unknown *Contributors*: PHenry, Sehome Bay

File:US 395.svg *Source*: http://en.wikipedia.org/w/index.php?title=File:US_395.svg *License*: unknown *Contributors*: HighwayMaster, Luokou, SPUI, 1 anonymous edits

File:I-15 (big).svg *Source*: http://en.wikipedia.org/w/index.php?title=File:I-15_(big).svg *License*: unknown *Contributors*: Ltljltlj, 1 anonymous edits

File:US 87.svg *Source*: http://en.wikipedia.org/w/index.php?title=File:US_87.svg *License*: unknown *Contributors*: Bidgee, SPUI, Xnatedawgx, 3 anonymous edits

File:US 83.svg *Source*: http://en.wikipedia.org/w/index.php?title=File:US_83.svg *License*: unknown *Contributors*: Bidgee, SPUI, Xnatedawgx, 2 anonymous edits

File:I-29.svg *Source*: http://en.wikipedia.org/w/index.php?title=File:I-29.svg *License*: unknown *Contributors*: Augiasstallputzer, Ltljltlj, SPUI, 1 anonymous edits

File:US 81.svg *Source*: http://en.wikipedia.org/w/index.php?title=File:US_81.svg *License*: unknown *Contributors*: Bidgee, Hellbus, SPUI, Xnatedawgx, 2 anonymous edits

File:I-35.svg *Source*: http://en.wikipedia.org/w/index.php?title=File:I-35.svg *License*: unknown *Contributors*: 1coolkid, Augiasstallputzer, Bidgee, Faiyazamzad, Fran Rogers, InterstateKazuya40, Kazuya35, Ltljltlj, SPUI, T2, Xnatedawgx, 1 anonymous edits

File:I-75.svg *Source*: http://en.wikipedia.org/w/index.php?title=File:I-75.svg *License*: unknown *Contributors*: Augiasstallputzer, Ltljltlj, SPUI, 1 anonymous edits

File:Business Loop 75.svg *Source*: http://en.wikipedia.org/w/index.php?title=File:Business_Loop_75.svg *License*: unknown *Contributors*: Engleman, I-215, KelleyCook, Ltljltlj, Mountain169257, SPUI, T2, 1 anonymous edits

File:US 11.svg *Source*: http://en.wikipedia.org/w/index.php?title=File:US_11.svg *License*: unknown *Contributors*: SPUI, 1 anonymous edits

File:US 7.svg *Source*: http://en.wikipedia.org/w/index.php?title=File:US_7.svg *License*: unknown *Contributors*: Rocket000, SPUI

File:I-95.svg *Source*: http://en.wikipedia.org/w/index.php?title=File:I-95.svg *License*: unknown *Contributors*: Juliancolton, Kazuya35, Ltljltlj, O, SPUI, 3 anonymous edits

File:NB 95.png *Source*: http://en.wikipedia.org/w/index.php?title=File:NB_95.png *License*: unknown *Contributors*: Myself (I formated the shield present on English Wikipedia)

Image:US Hwy 2 in Vermont.jpg *Source*: http://en.wikipedia.org/w/index.php?title=File:US_Hwy_2_in_Vermont.jpg *License*: unknown *Contributors*: User:Skeezix1000

Image:New England 15.svg *Source*: http://en.wikipedia.org/w/index.php?title=File:New_England_15.svg *License*: unknown *Contributors*: SPUI

Image:US blank.svg *Source*: http://en.wikipedia.org/w/index.php?title=File:US_blank.svg *License*: unknown *Contributors*: Fredddie, InterstateKazuya40, Martin H., Rfc1394, SPUI, Sehome Bay, 2 anonymous edits

Image:I-95 (ME).svg *Source*: http://en.wikipedia.org/w/index.php?title=File:I-95_(ME).svg *License*: unknown *Contributors*: Ltljltlj

Image:I-295 (ME).svg *Source*: http://en.wikipedia.org/w/index.php?title=File:I-295_(ME).svg *License*: unknown *Contributors*: Ltljltlj

Image:MA Route 100.svg *Source*: http://en.wikipedia.org/w/index.php?title=File:MA_Route_100.svg *License*: unknown *Contributors*: Rocket000, SPUI

Image:NB 161.png *Source*: http://en.wikipedia.org/w/index.php?title=File:NB_161.png *License*: unknown *Contributors*: User:Gordalmighty

Image:Northern Terminus of US Route 1.jpg *Source*: http://en.wikipedia.org/w/index.php?title=File:Northern_Terminus_of_US_Route_1.jpg *License*: unknown *Contributors*: User:Stratosphere

File:US 202 map.png *Source*: http://en.wikipedia.org/w/index.php?title=File:US_202_map.png *License*: unknown *Contributors*: User:25or6to4

File:US 13.svg *Source*: http://en.wikipedia.org/w/index.php?title=File:US_13.svg *License*: unknown *Contributors*: Bidgee, SPUI, 3 anonymous edits

File:US 40.svg *Source*: http://en.wikipedia.org/w/index.php?title=File:US_40.svg *License*: unknown *Contributors*: Rocket000, SPUI

File:Elongated circle 141.svg *Source*: http://en.wikipedia.org/w/index.php?title=File:Elongated_circle_141.svg *License*: unknown *Contributors*: Northenglish

File:I-76.svg *Source*: http://en.wikipedia.org/w/index.php?title=File:I-76.svg *License*: unknown *Contributors*: Augiasstallputzer, Fran Rogers, Ltljltlj, SPUI, 1 anonymous edits

File:Memorial to American Civil War veterans in Bethel, Maine.jpg *Source*:
http://en.wikipedia.org/w/index.php?title=File:Memorial_to_American_Civil_War_veterans_in_Bethel,_Maine.jpg *License*: unknown *Contributors*: Original uploader was Decumanus at en.wikipedia

Image:Main & Church Sts., Bethel, ME.jpg *Source*: http://en.wikipedia.org/w/index.php?title=File:Main_&_Church_Sts.,_Bethel,_ME.jpg *License*: unknown *Contributors*: Hugh Manatee

Image:Church Street, Bethel, ME.jpg *Source*: http://en.wikipedia.org/w/index.php?title=File:Church_Street,_Bethel,_ME.jpg *License*: unknown *Contributors*: Hugh Manatee

Image:The Prospect Hotel, Bethel, ME.jpg *Source*: http://en.wikipedia.org/w/index.php?title=File:The_Prospect_Hotel,_Bethel,_ME.jpg *License*: unknown *Contributors*: Hugh Manatee

Image:Ferry at West Bethel, ME.jpg *Source*: http://en.wikipedia.org/w/index.php?title=File:Ferry_at_West_Bethel,_ME.jpg *License*: unknown *Contributors*: Hugh Manatee

File:First Parish Church & Kennebunk Free Library, Kennebunk, ME.jpg *Source*:
http://en.wikipedia.org/w/index.php?title=File:First_Parish_Church_&_Kennebunk_Free_Library,_Kennebunk,_ME.jpg *License*: unknown *Contributors*: Photographer unknown

File:Seal of Kennebunk, Maine.jpg *Source*: http://en.wikipedia.org/w/index.php?title=File:Seal_of_Kennebunk,_Maine.jpg *License*: unknown *Contributors*: Canadaolympic989, TheDJ

File:Loudspeaker.svg *Source*: http://en.wikipedia.org/w/index.php?title=File:Loudspeaker.svg *License*: unknown *Contributors*: Bayo, Gmaxwell, Husky, Iamunknown, Mirithing, Myself488, Nethac DIU, Omegatron, Rocket000, The Evil IP address, Wouterhagens, 16 anonymous edits

File:AWikiElmA.jpg *Source*: http://en.wikipedia.org/w/index.php?title=File:AWikiElmA.jpg *License*: unknown *Contributors*: Original uploader was TJ aka Teej at en.wikipedia

File:View of Kennebunk Beach, ME.jpg *Source*: http://en.wikipedia.org/w/index.php?title=File:View_of_Kennebunk_Beach,_ME.jpg *License*: unknown *Contributors*: Photographer unknown

File:Old Storer Mansion, Kennebunk, ME.jpg *Source*: http://en.wikipedia.org/w/index.php?title=File:Old_Storer_Mansion,_Kennebunk,_ME.jpg *License*: unknown *Contributors*: Photographer unknown

File:Lexington Elms, Kennebunk, ME.jpg *Source*: http://en.wikipedia.org/w/index.php?title=File:Lexington_Elms,_Kennebunk,_ME.jpg *License*: unknown *Contributors*: Photographer unknown

Image:View of Kennebunk River 1903.jpg *Source*: http://en.wikipedia.org/w/index.php?title=File:View_of_Kennebunk_River_1903.jpg *License*: unknown *Contributors*: Boston & Maine Railroad.

Image:King's Highway, Kennebunk Beach, ME.jpg *Source*: http://en.wikipedia.org/w/index.php?title=File:King's_Highway,_Kennebunk_Beach,_ME.jpg *License*: unknown *Contributors*: Photographer unknown

File:Perkins Farm, West Kennebunk, ME.jpg *Source*: http://en.wikipedia.org/w/index.php?title=File:Perkins_Farm,_West_Kennebunk,_ME.jpg *License*: unknown *Contributors*: Photographer unknown

File:Troops Kennebunk Maine circa 1918.jpg *Source*: http://en.wikipedia.org/w/index.php?title=File:Troops_Kennebunk_Maine_circa_1918.jpg *License*: unknown *Contributors*: MarmadukePercy

File:Cumberlandcountyseal.JPG *Source*: http://en.wikipedia.org/w/index.php?title=File:Cumberlandcountyseal.JPG *License*: unknown *Contributors*: Cumberland County

File:Map of Maine highlighting Cumberland County.svg *Source*: http://en.wikipedia.org/w/index.php?title=File:Map_of_Maine_highlighting_Cumberland_County.svg *License*: unknown *Contributors*: User:Dbenbenn

Image:I-295.svg *Source*: http://en.wikipedia.org/w/index.php?title=File:I-295.svg *License*: unknown *Contributors*: Fran Rogers, Ltljltlj, SPUI

Image:MA Route 77.svg *Source*: http://en.wikipedia.org/w/index.php?title=File:MA_Route_77.svg *License*: unknown *Contributors*: SPUI

Printed by Books on Demand GmbH, Norderstedt / Germany